Robotics, Autonomous Systems and Contemporary International Security

Rapid technological advances in the field of robotics and autonomous systems (RAS) are transforming the international security environment and the conduct of contemporary conflict. Bringing together leading experts from across the globe, this book provides timely analysis on the current and future challenges associated with greater utilization of RAS by states, their militaries, and a host of non-state actors.

Technologically driven change in the international security environment can come about through the development of one significant technology, such as the atomic bomb. At other times, it results from several technologies maturing at roughly the same pace. This second image better reflects the rapid technological change that is taking us into the robotics age. Many of the chapters in this edited volume explore unresolved ethical, legal, and operational challenges that are only likely to become more complex as RAS technology matures. Though the precise ways in which the impact of autonomous systems – both physical and non-physical – will be felt in the long-run is hidden from us, attempting to anticipate the direction of travel remains an important undertaking and one that this book makes a critical effort to contend with.

The chapters in this book were originally published as a special issue of the journal *Small Wars & Insurgencies*.

Ash Rossiter is Assistant Professor of International Security at Khalifa University, UAE. His research rests at the nexus of technology and international security and he has published widely on this theme in journals such as *Parameters* and *Defence Studies*. He is author of the book *Security in the Gulf* (2020).

Robotics, Autonomous Systems and Contemporary International Security

Edited by
Ash Rossiter

Routledge
Taylor & Francis Group
LONDON AND NEW YORK

First published 2021
by Routledge
2 Park Square, Milton Park, Abingdon, Oxon OX14 4RN

and by Routledge
52 Vanderbilt Avenue, New York, NY 10017

Routledge is an imprint of the Taylor & Francis Group, an informa business

British Library Cataloguing in Publication Data
A catalogue record for this book is available from the British Library

ISBN 13: 978-0-367-62374-6

Typeset in Myriad Pro
by Newgen Publishing UK

Publisher's Note
The publisher accepts responsibility for any inconsistencies that may have arisen during the conversion of this book from journal articles to book chapters, namely the inclusion of journal terminology.

Disclaimer
Every effort has been made to contact copyright holders for their permission to reprint material in this book. The publishers would be grateful to hear from any copyright holder who is not here acknowledged and will undertake to rectify any errors or omissions in future editions of this book.

Contents

Citation Information

The chapters in this book were originally published in *Small Wars & Insurgencies*, volume 31, issue 4 (June 2020). When citing this material, please use the original page numbering for each article, as follows:

Introduction

The impact of robotics and autonomous systems (RAS) across the conflict spectrum
Ash Rossiter
Small Wars & Insurgencies, volume 31, issue 4 (June 2020), pp. 691–700

Chapter 1

Reluctant innovators? Inter-organizational conflict and the U.S.A.'s route to becoming a drone power
Marc R. DeVore
Small Wars & Insurgencies, volume 31, issue 4 (June 2020), pp. 701–729

Chapter 2

Armed, unmanned, and in high demand: the drivers behind combat drones proliferation in the Middle East
Francesco F. Milan and Aniseh Bassiri Tabrizi
Small Wars & Insurgencies, volume 31, issue 4 (June 2020), pp. 730–750

Chapter 3

U.S. drone campaign in Pakistan's Pashtun 'tribal' region: beginning of the end under President Trump?
Farooq Yousaf
Small Wars & Insurgencies, volume 31, issue 4 (June 2020), pp. 751–772

Chapter 4

What's in it for us? Armed drone strikes and the security of Somalia's Federal Government

Brendon J. Cannon
Small Wars & Insurgencies, volume 31, issue 4 (June 2020), pp. 773–800

Chapter 5
What's wrong with drones? Automatization and target selection
Andree-Anne Melancon
Small Wars & Insurgencies, volume 31, issue 4 (June 2020), pp. 801–821

Chapter 6
Friend or frenemy? The role of trust in human-machine teaming and lethal autonomous weapons systems
Aiden Warren and Alek Hillas
Small Wars & Insurgencies, volume 31, issue 4 (June 2020), pp. 822–850

Chapter 7
Bots on the ground: an impending UGV revolution in military affairs?
Ash Rossiter
Small Wars & Insurgencies, volume 31, issue 4 (June 2020), pp. 851–873

Chapter 8
Artificial intelligence, big data and autonomous systems along the belt and road: towards private security companies with Chinese characteristics?
Peter Layton
Small Wars & Insurgencies, volume 31, issue 4 (June 2020), pp. 874–897

Chapter 9
The impact of Artificial Intelligence on hybrid warfare
Guilong Yan
Small Wars & Insurgencies, volume 31, issue 4 (June 2020), pp. 898–917

Notes on Contributors

Brendon J. Cannon, Institute of International and Civil Security, Khalifa University, Abu Dhabi, UAE.

Marc R. DeVore, School of International Relations, University of St. Andrews, UK.

Alek Hillas, School of Global, Urban and Social Studies, RMIT University, Melbourne, Australia.

Peter Layton, Griffith Asia Institute, Griffith University, Brisbane, Australia.

Andree-Anne Melancon, Department of Defence and International Affairs, Royal Military Academy Sandhurst, UK.

Francesco F. Milan, Department of Defence Studies, King's College London, UK.

Ash Rossiter, Institute of International and Civil Security, Khalifa University, Abu Dhabi, UAE.

Aniseh Bassiri Tabrizi, International Security Studies, Royal United Services Institute, London, UK.

Aiden Warren, School of Global, Urban and Social Studies, RMIT University, Melbourne, Australia.

Guilong Yan, Foreign Military Studies Centre, PLA Strategic Support Force Information Engineering University, Henan, China.

Farooq Yousaf, Department of Politics, University of Newcastle, Callaghan, Australia.

The impact of robotics and autonomous systems (RAS) across the conflict spectrum

Ash Rossiter

RAS-driven revolution?

Technology is often held up to be a key – if not *the* key, for some at least – driver in shaping the character or even perhaps the nature of war. The introduction of new technologies has resulted in novel and, usually by implication, more effective ways of *conducting* warfare. Indeed, many leading military thinkers have identified new technology – both of civilian and military origins – as the source of many of history's military revolutions, or so-called major military innovations (MMIs), as others have termed radical change in military affairs.[1] Whilst technological change has not accompanied all prior MMIs, it has nonetheless been the root cause of many. And because those able to exploit technologically driven MMIs are, all else being equal, at an advantage over those who are unable to, technology can be said to have played a leading role in changes to the global balance of military power.[2]

This is not the only reason why technology is of great interest to scholars of international relations (IR) and military affairs. New technology also possesses the potential to reshape the international security environment below the level of state relations. At different points in history, technological change has recalibrated the relative power equation between state authorities and non-state elements. Indeed, various technological developments in recent decades have arguably enhanced the ability of non-state armed groups to challenge state authorities. Historically, the actions available to militants have been constrained by lack of access, relative to states, to advanced weapons and equipment; non-state groups, including armed militants, can now access capabilities that were, until recently, largely within the purview of governments. If states can no longer bank on a monopoly over advanced technologies, the playing field of conflict may be evening somewhat.[3] Much has been written, for example, about how access to forms of readily available communication and multi-media technology has revolutionized the way a variety of groups raise funds, recruit supporters and fighters, conduct operations, and publicize activities, especially successful attacks.[4]

Technologically driven change can come about through the adoption of one significant technology, such as the combustion engine. At other times, it might be caused by several technologies maturing and converging at roughly the same time. Here it is the synergy between complementary technologies which contains the revolutionary effect rather than the revolutionary potential of any one of the contributing technologies on its own.

For many, this second image of technological change more accurately reflects the rapid changes afoot; change that is taking us into a new epoch: the robotics age. Proponents of the idea that we are the cusp of a robotics revolution base their prognosis not so much on developments in robotics hardware (i.e. the mechanical platforms and their physical sub-components) but on the machines software, allowing them to possess greater intelligence and perform functions with ever greater degrees of autonomy. In this way, the field of robotics and autonomous systems (RAS) can be described as covering developments across a wide range of inter-connected technologies including un-crewed ground systems, aerial systems, AI, and self-learning machines. RAS is an accepted term in academia and the science and technology (S&T) community and highlights the physical (robotic) and cognitive (autonomous) aspects of these systems.

Military robotic systems, especially those machines with increasing levels of autonomy, are already making their influence felt in international affairs in myriad ways. The most clear-cut example is of course the use of unmanned aerial vehicles (UAVs) carrying lethal payloads and used for targeted killings. Reflecting the continued importance of lethal UAV usage for international security, this special issue carries several contributions on this topic. But it is perhaps the latent potential of *autonomous* systems that has stoked greatest interest. As one leading thinker about the future impact of RAS writes:

> Autonomous Systems raise challenging operational, strategic, and policy issues, the full scope of which cannot yet be seen. The nations and militaries that see the furthest into a dim and uncertain future to anticipate these challenges and prepare for them now will be best poised to succeed in the warfighting regime to come.[5]

Deep questions remain, however, as to who is set to benefit – or, more to the point, will be disadvantaged – when the promise of RAS for military use is fully realized. For some, the underlying assumption is that only the most powerful nations and their militaries can benefit from higher-end RAS capabilities. In contrast, others see RAS as a democratizing technology, that is, small states and even non-state actors will have access to a suite of technologies almost comparable to that of the major powers. The rate at which RAS-related technology proliferates and the level of difficulty involved in adopting it, will be an important factor.[6]

Military robotics: the state of play

Many advanced economies have been converting their expertise in private-sector robotics into new military investments for some years. Operationally, military robotic systems have been used primarily in missions where they substitute for inhabited platforms. Many states have used UAVs, for example, in the place of manned aircraft for both surveillance and strike missions. In a similar vein, the US extensively used unmanned ground vehicles (UGVs) in Iraq and Afghanistan (and is still doing so) in EOD (explosive ordinance disposal) roles. To summarize the main motive behind this substitution would be to say that military robotics are a way of putting the focus of warfare closer to the enemy but at the same farther from oneself.[7] Or, to put another way, it is the desire to place metal before flesh.

The incorporation of military robotics has mostly been a 1:1 replacement of traditionally manned systems or of a human-performed task, such as EOD. But such systems not only have the potential to perform conventional dirty, dull, and dangerous military tasks, such as surveillance and counter mining; they may also change the way military operations are conducted in the future. They might even make it possible to envisage new types of mission altogether. In doing so, they would move beyond being considered a sustaining technology, enhancing capabilities along established lines of effort; instead, military robotics could claim to be a revolutionary technology. Much of the excitement with military robotics lies in the future direction of moving from one-to-one substitution to many-to-one replacements. In place of large capital-intensive manned platforms would be smaller, cheaper and networked unmanned systems. The appeal is not only lower unit costs but the dispersion of risk.

Much care should be taken, however, in attempts to predict the near-term revolutionary potential of any given technology. So-called game-changing technologies often fall well short of expectations. Indeed, exuberance or hype about any given technology can lead to unrealistic expectations about its immediate impact.[8] To date, the vast majority of military robotic systems are remotely controlled/piloted or, at the very least, remotely supervised. Military robotics in the shape of UAVs have excelled in relatively benign airspace where their inherent weaknesses in maneuverability and survivability have not so far outweighed their key strength: endurance.

In comparison to UAVs, military robotics in other domains – land, sea, and underwater – have not had anywhere near the same impact. There are additional military effectiveness issues with military robotics operating in these domains, especially the complexity of the environments they need to operate in. Moreover, a concern relevant to all military robotics regardless of where they operate is information security. As Michael Horowitz points out, 'Military robotics require not just large amounts of bandwidth, but secure access to that bandwidth.'[9] This makes them especially vulnerable to hacking. Insurgents in Iraq reportedly hacked into MQ-9 Reaper drones in 2008 and 2009.[10] Most importantly, the potential loss of connectivity raises serious questions about networked military robotics. Severely degraded or denied access to information systems could render many military robotics systems irrelevant, because operators would lose access to those systems. The silver bullet solution, of course, is increasing the level of autonomy of these systems to reduce dependency on communication within and without their network.[11] The future generation of military robotics will likely see the relinquishment of some human control in the pursuit of effectiveness.

Towards greater autonomy

Simply put, autonomy is the ability of a machine to perform a task without human input. Under this definition, an autonomous system is a machine, whether hardware or software, that once activated performs some task or function on its own.[12] Autonomy can be described as the cognitive engine that powers robots. There are some existing autonomous weapon systems, such as the radar-guided and computer-controlled Phalanx gun, a close-in weapon system (CIWS) for defense against airborne threats such as anti-ship missiles. These have autonomous settings but follow pre-programmed actions within tightly set parameters in highly controlled environments.

Some weapon systems are already configured so humans are out of the decision-making process. This is usually for situations when there is not enough time for humans to react. This is for defensive purposes against non-human targets, such as against missiles and mortars [what about drones]. According to a US DOD Directive: 'Human-supervised autonomous weapon systems may be used to select and engage targets, with the exception of selecting humans as targets, for local defense to intercept attempted time-critical or saturation attacks for: (a) Static defense of manned installations. (b) Onboard defense of manned platforms.'[13]

In this way, they might better be described as *automatic* rather than the sorts of autonomous systems under development that can potentially be used in unstructured environments. The emerging technology is likely to enhance performance and capability options available to militaries and other users through automatic and autonomic control before full autonomy will be available.

Autonomy in weapon systems has been under development and controversy for many years. But rapid advances in AI are transforming the capability of a computer systems to perform tasks that normally require human intelligence. These advances in AI – whether developed in computer software or physical hardware – will be transformative. The increasing ability of artificial systems to perform tasks that require human-like perception, cognition, planning, learning, communication, or physical action endows them with revolutionary potential, especially if these machines can target and fire by themselves. This is not to say that lethal autonomous weapon systems (LAWS) are inevitable, but the proliferation of proliferation of RAS and AI internationally, ranging from foreign militaries, to violent non-state groups to criminal organizations, does make their eventual appearance all the more likely.

Droning on

To date, when commentators write about military robotics, they typically think about one subset of this category – unmanned aerial vehicles, or drones. This is unsurprising, given the importance of armed drone campaigns as part of the so-called war on terror. Drones may act as long-endurance communication relays (supporting the network), as small, inexpensive, fast-moving, hard-to-detect sensors (nodes to support a common operating picture), and as platforms for delivering precision strikes against targets that are too difficult or dangerous for manned platforms to reach.[14] It is this latter category (i.e. the armed varieties that have been used for strike campaigns) which are most discussed. Drones' current preeminence in a variety of military operations leads many today to assume that their development was straightforward; teleologically determined by technological advances. But drones' convoluted evolution raises questions about the military innovation process. The opening article in this special issue by Marc DeVore shows how in the case of the US, the Air Force's and naval aviation's pilot-dominated hierarchies never prioritized drones of their own accord because of their preferences for manned aircraft. Political leaders, meanwhile, lacked the expertise to judge what drone technologies could achieve and therefore could not compel the military to embrace them. It was, within this context, competition from organizations – the CIA, the Navy's surface warfare community and the Army – that did not hold the preservation of manned aviation as a core belief that obliged reluctant aviators to embrace drones.

Whatever reluctance that may have existed among some of the American services towards drones, most military and political leaders today see drones as an essential capability to posses. Francesco Milan and Aniseh Tabrizi look at how combat drones (unmanned combat aerial vehicles, or UCAVs) have become key strategic enablers for state actors in the Middle East. For them, UCAVs are no longer seen as an optional asset. With the development of a multitude of indigenous UCAV projects, and the arrival of Chinese-made armed drones on the international market, military procurement in the Middle East has entered a new phase, in which possessing armed drone capabilities is becoming the norm. The paper adds significantly to the literature on drone proliferation.[15]

The next couple of articles provide new perspectives on two ongoing drone campaigns. Farooq Yousaf looks again at what seems for many analysts to be the coming end of the long-running US drone campaign in Pakistan's Pashtun 'tribal' areas. But as Yousaf argues, although the last drone strike took place almost a year ago, it does not mean that the 'tribal' areas have witnessed an end to the campaign. If high-profile terrorists, based on actionable intelligence, are located in the region, the US will readily strike again.

Brendon J. Cannon provides a counterpoint to previous studies of drone campaigns by measuring the Federal Government of Somalia's (FGS) reliance on the US drone campaign in Somalia in order to maintain its tenuous hold on power. It measures the extent to which the drone campaigns of external parties are able to insulate weak host governments from the threat of an internal foe. It finds that US drone strikes offer the occasional but only reliable check on al-Shabaab thereby allowing the FGS to continue functioning in Mogadishu.

Where the rubber hits the road

All new military technologies face initial periods of uncertainty regarding their effectiveness and pose organization issues surrounding their employment.[16] The following series of articles look at enduring ethical, practical, and technological challenges that governments and military organizations face in operationalizing increasingly autonomous machines. In particular, delegating more control to machines is a trade-off, and one with important moral, ethical, and military implications.

Andree-Anne Melancon argues that before autonomous attack decisions can be debated, issues related to target selection need to be considered. Focusing the discussion on drones and artificial intelligence, she argues, overshadows more fundamental questions. The article's main argument is that the main ethical problem raised by autonomous drones do not come from the weapon system or technology itself. Instead, they stem from the challenges of target identification. Thus, asking 'what's wrong with drones?' is for Melancon misleading as it focusses on the technology instead of the primary issue of target selection.

In 'Friend or Frenemy?', Aiden Warren and Alek Hillas explore the imprecise boundary between Lethal Autonomous Weapons Systems (LAWS) and Human-

Machine Teaming – as a subset of Human-Machine Interaction – and the extent both are emerging as a point of concern (and option) in military and security policy debates. As the development of Human-Machine Teaming relates to artificial intelligence (AI) capabilities there also exists an area of concern pertaining to reliability and confidence, particularly in the heat of battle. Also known as Manned-Unmanned Teaming, Human-Machine Teaming attempts to engender trust and collaborative partnerships with robots and algorithms. Clearly the prospect of LAWS in recent times, or so-called 'killer robots,' has raised questions relating to the degree such devices can be trusted to select and engage targets without further human intervention.

Ash Rossiter's article on unmanned ground vehicles (UGVs) argues that when it comes to the operational use of such systems – armed or otherwise – their recent history can best be described as one of promise so far unfulfilled. By tracing past and present efforts to develop and field UGVs – and the enduring challenges that lie therein – this article attempts to gauge the likely impact of such systems in future conflict, as well as their effect on international security more broadly. The article concludes that although UGVs will almost certainly become a major – if not indispensable – feature of future military land operations, they will, similar to other promising militarily relevant technologies before them, continue to produce unrealistic expectations about their impending revolutionary effect.

AI's promise

The final two articles provide unique perspectives on the application of AI. Peter Layton looks at how the Chinese state is undertaking a major 'anti-secession and counter-terrorism' campaign in Xinjiang using a variety of high-technology means: artificial intelligence, big data, wireless connectivity, autonomous systems, and robotics. For Layton, the demand and supply sides seem to be in sync, and believes Chinese private security companies will soon use a suite of advanced information technology systems with a proven employment doctrine across much of the Belt and Road Initiative (BRI) and beyond. If the Chinese Communist Party seeks 'socialism with Chinese characteristics,' so China's new digitized and automated PSCs may become considered 'PSCs with Chinese characteristics.'

In 'The Impact of Artificial Intelligence on Hybrid Warfare,' Guilong Yan discusses the impact of Artificial Intelligence on the five instruments of power – military, political, economic, civil, and informational (MPECI), and analyses the changes and continuities of hybrid warfare in the age of Artificial Intelligence. In the future, according to Yan, HW, innovative and fresh-new means of fighting enabled by AI will surely come out, and will be used to achieve tactical or even operational surprise. However, the limitations of human imagination do not prevent us from drawing the conclusion that

HW in the AI age will look very different such that the kinetic aspect of war will become more lethal, precise, and fast-tempo, and that the competition for control and dominance of data and information will be of vital importance.

Looking ahead

Many of the above essays have looked a pre-existing and unresolved issues that are likely to become more complicated as RAS technologies develop. Though the precise ways in which the impact of autonomous systems – both physical and non-physical – will be felt in the long-run is hidden from us, the task of attempting to predict the emerging direction remains nonetheless an important one. Growing technological overlap between rapidly expanding commercial use of AI and robotics, and the accelerating 'Spin in' of these technologies into the military realm, make their operational use a possibility in the near future.[17] But will actors have comparable access to such technology or will emerging RAS technologies favor the powerful over the weak, allowing the former to surge ahead of the latter in terms of capabilities? One thing is for sure – and this is a common theme binding many of the contributing articles together – the development of militarily useful RAS-related technologies and their adoption by a range of actors will continue to throw up major challenges. This special issue, with its diverse set of contributions, is an attempt to take forward this discussion in the knowledge that much still needs to be done now and in the future to better understand and anticipate the impact of RAS across the conflict spectrum.[18]

Notes

1. Gray, *Strategy for Chaos*; Boot, *War Made New*; Krepinevich, "Cavalry to Computer"; Van Creveld, *The Transformation of War*; and Mahnken, *Technology and the American Way of War*. Though it should be added, Clausewitz has very little to say about the technology's impact on war.
2. RMAs or MMIs can also of course be driven by major organizational changes to militaries. See Horowitz, *The Diffusion of Military Power*, 22.
3. Rossiter, "Participation in Warfare."
4. Diamond, "Liberation Technology."
5. Scharre, "The Opportunities and Challenges."
6. For a synopsis of the major arguments related to horizontal diffusion of military technology, see Gilli and Gilli, "Why China has Not Caught Up Yet."
7. Kline, "Impacts of the Robotics Age," 67.
8. For a detailed examination of how hype can influence the development cycle of technology, see Rossiter, "High-Energy Laser Weapons."
9. Horowitz, "Military Robotics, Autonomous Systems," 182.
10. MacAskill, "US Drones Hacked."
11. Horowitz, "Military Robotics, Autonomous Systems," 162.

12. Scharre, "The Opportunity and Challenges."
13. US DOD, "Autonomy in Weapon Systems."
14. Dombrowski, Gholz, and Ross, "Selling Military Transformation," 530.
15. Horowitz, Kreps, and Fuhrmann, "Separating Fact from Fiction"; Fuhrmann and Horowitz, "Droning On"; Boyle et al. "Debating Drone Proliferation"; and Joshi and Stein, 'Emerging Drone Nations," 53–78.
16. This is an idea throughout Horowitz's *The Diffusion of Military Power*.
17. Altmann Frank Sauer, "Autonomous Weapon Systems," 117.
18. See Rossiter, "Drone Usage by Militant Groups."

Disclosure statement

No potential conflict of interest was reported by the author(s).

ORCID

Ash Rossiter http://orcid.org/0000-0001-5427-2229

Bibliography

Altmann, Jürgen, and Frank Sauer. "Autonomous Weapon Systems and Strategic Stability." *Survival* 59, no. 5 (2017): 117–142. doi:10.1080/00396338.2017.1375263.

Boot, Max. *War Made New: Technology, Warfare, and the Course of History: 1500 to Today*. New York: Gotham Books, 2006.

Boyle, Michael J., Michael C. Horowitz, Sarah E. Kreps, and Matthew Fuhrmann. "Debating Drone Proliferation." *International Security* 42, no. 3 (2018): 178–182. doi:10.1162/ISEC_c_00308.

Diamond, Larry. "Liberation Technology." *Journal of Democracy* 21, no. 3 (2010): 69–83. doi:10.1353/jod.0.0190.

Dombrowski, Peter J., Eugene Gholz, and Andrew L. Ross. "Selling Military Transformation: The Defense Industry and Innovation." *Orbis* 46, no. 3 (2002): 523–536. doi:10.1016/S0030-4387(02)00122-9.

Fuhrmann, Matthew, and Michael C. Horowitz. "Droning On: Explaining the Proliferation of Unmanned Aerial Vehicles." *International Organization* 71, no. 2 (2017): 397–418. doi:10.1017/S0020818317000121.

Gray, Colin S. *Strategy for Chaos: Revolutions in Military Affairs and the Evidence of History*. London: Frank Cass, 2002.

Gilli, Andrea, and Mauro Gilli. "Why China has not Caught Up Yet: Military-Technological Superiority and the Limits of Imitation, Reverse Engineering, and Cyber Espionage." International Security 43, no. 3 (2018–19): 141–189.
Horowitz, Michael C. *The Diffusion of Military Power: Causes and Consequences*. Princeton, NJ: Princeton, 2010.
Horowitz, Michael C. "Military Robotics, Autonomous Systems, and the Future of Military Effectiveness." In *The Sword's Other Edge: Trade-offs in the Pursuit of Military Effectiveness*, edited by Dan Reiter, 161–196. Cambridge, UK: Cambridge University Press, 2017.
Horowitz, Michael C., Sarah E. Kreps, and Matthew Fuhrmann. "Separating Fact from Fiction in the Debate over Drone Proliferation." *International Security* 41, no. 2 (2016): 7–42. doi:10.1162/ISEC_a_00257.
Joshi, Shashank, and Aaron Stein. "Emerging Drone Nations." *Survival* 55, no. 5 (2013): 53–78. doi:10.1080/00396338.2013.841805.
Kline, Jeffrey E. "Impacts of the Robotics Age on Naval Force Design, Effectiveness, and Acquisition." *Naval War College Review* 70, no. 3 (2017): 63–77.
Krepinevich, Andrew. "Cavalry to Computer: The Pattern of Military Revolutions." *The National Interest* 37 (1994): 30–43.
MacAskill, Ewan. 2009. US Drones Hacked by Iraqi Insurgents. *The Guardian*, December 19.
Mahnken, Thomas G. *Technology and the American Way of War since 1945*. New York: Columbia University Press, 2008.
Rossiter, Ash. "Drone Usage by Militant Groups: Exploring Variation in Adoption." *Defense & Security Analysis* 34, no. 2 (2018): 113–126. doi:10.1080/14751798.2018.1478183.
Rossiter, Ash. "High-Energy Laser Weapons: Overpromising Readiness." *Parameters*, 48, no. 4 (2018): 35–46.
Rossiter, Ash. "Participation in Warfare: Are We Witnessing Paradigmatic Change?" *Journal of Military and Strategic Studies* 19, no. 3 (2019): 103–123.
Scharre, Paul D. 2017. "The Opportunities and Challenges of Autonomous Systems." Conference Paper for Autonomous Systems, Issues for Policymakers, HQ SACT, NATO, Norfolk, Virginia.
US Department of Defense [DOD]. 2012. "Autonomy in Weapon Systems." *Directive no. 3000.09*, November 21. http://www.esd.whs.mil/Portals/54/Documents/DD/issuances/dodd/300009p.pdf.
Van Creveld, Martin. *The Transformation of War*. New York: Free Press, 1991.

Reluctant innovators? Inter-organizational conflict and the U.S.A.'s route to becoming a drone power

Marc R. DeVore

ABSTRACT
Few innovations have marked the late-20th and early-21st centuries more than unmanned aerial vehicles, also known as drones. Drones' current preeminence leads many to assume that their development was teleologically determined by technological advances. The empirical record, however, belies such assumptions and is filled with vicissitudes. The Air Force's and Naval aviation's pilot-dominated hierarchies never prioritized drones over manned aircraft of their own accord. Politicians, meanwhile, lacked the expertise to judge what technologies could achieve and therefore could not compel the military to embrace drones. It was, thus, competition from other organizations – the CIA, the Navy's surface warfare community and the Army –that obliged reluctant aviators to embrace drones. My study's key original finding is that inter-agency competition impels militaries to embrace technologies that they would otherwise reject. Warfare's evolution means that non-military bodies – intelligence agencies, interior ministries and paramilitary forces – develop capabilities that rival those of traditional military services in specific domains and these organizations can prove more agile at adopting certain new technologies because of their flatter organizational structures.

Introduction

Few military innovations have marked the late-20th and early-21st centuries more than unmanned aerial vehicles, which are better known as drones. Long considered a marginal military tool, drones have first become central to great powers' visions of future warfare. Predator and Reaper drones have spearheaded the United States' campaign to kill insurgents and terrorist leaders. American military planners, meanwhile, envisage drones replacing manned aircraft and satellites for such missions as reconnaissance from the stratosphere, communications relays and mid-air refueling. All major powers even now are experimenting with stealth drones designed from air-to-air combat and attacking heavily defended targets. Many experts anticipate that drones will soon supplant manned combat aircraft in the foreseeable future given the meteoric development of artificial intelligence.

Drones' current preeminence leads many today to assume that their development was straightforward; teleologically determined by technological advances. The empirical record, however, belies such assumptions and is rather filled with vicissitudes. The United States Air Force, which is today the world's premier drone user, actually developed a sophisticated drone arsenal in the 1960s, which it used to collect intelligence on and conduct airstrikes against North Vietnam's most heavily defended targets.

After the Vietnam War, however, the Air Force disbanded and scrapped its drone force. A decade then elapsed before the Navy reintroduced drones into front-line American service, by purchasing inexpensive Israeli drones to operate from antiquated American battleships. After the Navy brought back drones in the 1980s, it was the CIA that developed the Predator drone's forebearer – the Gnat 750 – in the 1990s. Only at the end of this process, in the late-1990s, did the Air Force once again embrace drones, thereby setting the stage for the dramatic growth in their usage during the so-called War on Terror.

Drones' convoluted evolution raises questions about the military innovation process. Why, for example, did the Air Force marginalize drones even after their value had been proven in combat? The corollary to this question is why did the Air Force's leadership warmly embrace drones during certain periods, while it rejected them during others? To answer these questions, I assess the United States' development of drones up until they achieved prominence after the September 2001 terrorist attacks.

To preview my conclusions, the case of drones highlights the central role of inter-organization competition in military innovation. The Air Force's and Naval aviation's pilot-dominated hierarchies never prioritized drones of their own accord because of their preferences for manned aircraft. Political leaders, meanwhile, lacked the expertise to judge what drone technologies could achieve and therefore could not compel the military to embrace them. It was, within this context, competition from organizations – the CIA, the Navy's surface warfare community and the Army – that did not hold the preservation of manned aviation as a core belief that obliged reluctant aviators to embrace drones.

My study's key original finding is that inter-agency competition impels military organizations to embrace technologies that they would otherwise reject. Warfare's evolution means that non-military bodies – intelligence agencies, interior ministries and paramilitary police forces – develop capabilities that render them rivals of traditional military services in specific domains. Since such organizations feature different career trajectories and hierarchical structures from military services, they can prove more agile at adopting new technologies.

Agents of innovation

The over-riding importance of new ways of waging war – armoured divisions, aircraft carriers and radar-directed air defence – stirred many military professionals to regard the timely adoption of new technologies as crucial to success following the Second World War. French General Charles Ailleret best summarized this new reality when he wrote, 'the greatest change in the art of war is that a military [today] is bound to be crushed if it does not apply itself intensely to ensuring that its weapons are both "up to date" and adapted to its future missions.'[1] Nuclear weapons' and long-range missiles' impact on the public consciousness in the years following the Second World War then drove scholars to first systematically examine how states innovate militarily. Despite the ensuing decades of academic inquiry, experts still disagree as to how military innovation occurs.

Political leaders constitute the first category of actor whom certain scholars argue are central to innovation. According to Barry Posen, political leaders force military organizations to embrace new technologies, tactics and modes of organization.[2] Military organizations, within this context, are conceptualized as parochial organizations whose ability to embrace new means of making war is hindered by their bureaucratic politics. Innovation, consequently, occurs when the international environment alarms politicians to such an extent that they force their militaries to embrace innovations essential to national security. Political leaders, according to Posen, accomplish this by partnering with 'mavericks' within the military, whose ideas are otherwise ignored by higher ranking officers.[3]

Although Posen offers a particular vision of how political leaders promote innovation, other scholars support his general message. Kimberley Marten Zisk and Sten Rynning, for example, detail how leaders can build up and then draw upon think tanks filled with civilian national security experts to develop and impose new military doctrines.[4] Studying the Soviet case, Matthew Evangelista goes further, arguing that Soviet institutions were so hierarchical that only political leaders' intervention could generate meaningful change, such as when Premier Nikita Khrushchev reallocated resources from conventional platforms to missiles.[5]

An alternative school of thought argues that military organizations develop innovations endogenously. Stephen Rosen led the way in this regard, arguing that political leaders lack the specialist knowledge needed to promote innovations and that military organizations are not as conservative as Posen asserted.[6] Military officers, according to Rosen, possess more comprehensive understandings of tactical trends and enemy capabilities than their political superiors. Innovation therefore occurs when top-level commanders recognize emerging trends in war-making. Military leaders, within this context, promote junior officers in new specialities and thereby lay the basis for gradual, but critical innovations.

Recent scholarship on the Second World War supports this re-interpretation of classic military innovations. New studies of Germany's blitzkrieg,[7] Japanese aircraft carriers[8] and American naval preparations for the Pacific War[9] all highlighted the role of highly functioning, yet staid military bureaucracies in generating revolutionary innovations. Moreover, historians first and then political scientists recognized military organizations' ability to generate innovations from the bottom up. Battlefield experience, whether in World War I trenches[10] or Normandy's hedgerows,[11] drove military organizations to rapidly re-evaluate and improve how they waged war.

Not all scholars, however, proved so sanguine as to military bureaucracies' aptitude to reform themselves. Many scholars, indeed, view armed forces' ingrained platform communities as an obstacle to their embracing radical innovations. Edward Katzenbach's seminal study of the survival of the horse cavalry units until the mid-20th century forcefully made this point.[12] According to Katzenbach, the fact that top cavalry officers had spent their entire careers mastering a particular form of warfare burdened them with motivated biases when it came to evaluating their branch's future. According to Katzenbach, 'Romanticism, while perhaps stultifying realistic thought, gives a man that belief in the value of the system he is operating that is so necessary to his willingness to use it in battle …. But faith [in a weapons system or tactic] breeds distrust of change.'[13]

Building on Katzenbach's insight, Carl Builder argues that military services will embrace only those new weapons that enhance their dominant platform communities, while rejecting those that do not.[14] Elizabeth Kier and David Johnson, in related arguments, postulate that military services' distinctive organizational cultures often prevent them from adopting the optimal structures and tactics.[15] Other scholars, meanwhile, argued that armed services' self-conceptions can lead them to reject superior ways of waging war altogether.[16]

Examined holistically, neither of the two models detailed above – of civilian-led and military-led innovation – is wholly convincing. While political leaders, indeed, have incentives to force their armed forces to adopt the most innovative and effective ways of waging war, they generally lack the expertise and managerial leverage to do so. Robert Coulam's study of American political leaders' unsuccessful efforts to shape combat Cold War aircraft projects bolsters this finding.[17] Military bureaucracies, however, after often loath to embrace radical changes that would threaten their dominant platform communities. If political leaders lack the wherewithal to promote innovations and military commanders will hesitate to enact reforms that could undermine their cherished combat arms, then the question remains as to how military innovation occurs?

Inter-organizational competition

One mechanism whereby otherwise reluctant military organizations can be prompted to adopt radical innovations lies in inter-organizational conflict.

Armed services, indeed, regularly compete with one another and with other government agency for their roles and missions. Achieving primary responsibility for fulfilling key missions, within this context, increases the resources that an organization will command and enhances its prestige at the national level. This incentive to secure leadership over important national security missions, and the accompanying fear of losing responsibility to a rival organization, provides a powerful incentive for armed services to adopt even those military innovations they would otherwise eschew. Two varieties of inter-organizational competition – inter-service and inter-agency – can produce such an impetus to innovation.

Policymakers have traditionally regarded inter-service rivalry as an impediment to the sound management of a state's defence. Such rivalries are often viewed as needlessly parochial, diverting resources from the efficient accomplishment of objectives. Often referenced cases supported this point-of-view. President Franklin Roosevelt, for example, divided the Pacific Ocean into two operational theatres during World War II to appease the United States' Army and Navy, which each wanted to control the war against Japan. Japan's Army and Navy, for their part, wasted resources duplicating one another's capabilities, with the Army acquiring aircraft carriers and the Navy possessing its own infantry. Michael Armacost's seminal study of Cold War missile programs, however, offers an alternative view of inter-service rivalry.[18]

Armacost's examination of the United States Army's and Air Force's ballistic missile programs led him to conclude that inter-service competition drives innovation. The Air Force, within this context, accorded a low priority to ballistic missiles until the Army launched its competing Jupiter missile program. Rivalry between two design teams, beholden to separate services, thereafter spurred technical advances that improved both missile programs, such as when the Army's superior heat-shielding for warheads was adopted by both services. Armacost's finding – that inter-service competition fuels innovation – was unambiguous.

While Armacost induced the relationship between inter-service competition and innovation from his study, it fell to Owen Coté to theorize the relationship.[19] According to Coté, inter-service rivalry is particularly prone to foster innovation because it forces military services to embrace even those changes that threaten existing platform communities because doing so will be less prejudicial than allowing a rival service to dominate the new capability. In Coté's own words, inter-service competition creates conditions where 'a service becomes interested in developing a new combat arms even when it comes at the expense of existing ones, in order to fend off the claims of a branch of a rival service seeking a monopoly on the new instrument of war.'[20]

Ultimately, Coté argues that such inter-service competition indirectly empowers political leaders. Such is the case because 'the central authority responsible for the adjudication of interservice arguments and negotiations over doctrine is ultimately a civilian one.'[21] In other words, inter-service competition forces military services to appeal to political leaders, couching their rival arguments in terms comprehensible to civilian policymakers and allowing these later to decide which services' claims to support.

While Armacost and Coté persuasively argue that inter-service competition promotes military innovation, they neglect the possibility that competition between military services and other governmental agencies can drive military innovation as well. Powerful reasons exist, however, for anticipating that inter-service competition can play an analogous role to inter-service rivalry in this regard. The evolving security challenge facing states, indeed, has led non-military agencies to acquire capabilities that rival those of armed services in specific domains. Intelligence agencies, for example, can develop reconnaissance aircraft, surveillance satellites and paramilitary units. Coastguards, meanwhile, procure fleets of patrol vessels, helicopters and surveillance aircraft that can enforce maritime claims more efficiently than navies. Law enforcement agencies, likewise often develop counter-terrorist and riot control capabilities that can efficaciously operate abroad.

Since such agencies can technically compete with armed services for specific roles and missions, they can also logically spur the latter to innovate. Such is particularly the case because civilian agencies possess comparative advantages when it comes to innovating. A considerable management literature contends that small and agile corporations often out-innovate larger industrial behemoths. Since civilian agencies – intelligence, coastguards and law enforcement – generally possess flatter organizational structures and more flexible personnel policies than armed services, they should demonstrate analogous advantages when it comes to embracing new ways of operating.

Considerable anecdotal evidence suggests that such inter-agency competition can drive military innovation. It was the CIA's Science and Technology Directorate, for example, that compelled the Air Force to embrace stealthy aircraft and reconnaissance satellites.[22] In both cases, the CIA's more flexible procurement procedures and greater solicitude for acquiring high quality imagery led it to invest in risky technologies in which the Air Force had demonstrated little interest. The Air Force, by way of contrast, viewed reconnaissance as a low-priority mission and uncreatively modified conventional combat aircraft for this highly demanding mission, which ultimately proved an inadequate approach. Indeed, the Air Force only sought to acquire specialized reconnaissance aircraft and imagery satellites after the CIA demonstrated their potential.

In a like manner, France's paramilitary police – the Gendarmerie – developed expeditionary counter-terrorism capabilities that later impelled the armed forces to invest in their own special forces. In this case, the Gendarmerie seized upon the rise in international terrorism in the 1970s to develop counter-terrorism units, the *Groupe d'Intervention de la Gendarmerie Nationale* (GIGN) and *Escadron Parachutiste d'Intervention de la Gendarmerie Nationale* (EPIGN).[23] These forces' specialized skills led political leaders to regard them as a favoured tool for discreet military operations. They, for example, deployed the EPIGN to protect French diplomats in conflict zones ranging from Chad to Lebanon.[24] More spectacularly, they 'lent' the GIGN to Saudi Arabia to help retake Mecca's Grand Mosque from terrorists in 1979 and ordered it to rescue hostages held in caves from the grip of pro-independence terrorists in New Caledonia in 1988. Impelled by this threat to their own role in low-intensity military operations, France's Joint General Staff formed a special operations command (the *Commandement des Opérations Spéciales*) in 1992.

Inter-service and inter-agency competition thus offer two potential mechanisms for obliging reluctant armed services to innovate. These should logically play a significant role with drones since the armed service best positioned to play the leading role – the Air Force – is also likely to be the most reticent to do so because of its domination by combat aircraft pilots, who regard drones as a threat to their platform community. In the following pages I will comparatively test the extent to which this innovation was driven by civilian leaders, military bureaucracies, inter-service rivalry and interagency competition. To this end, I adopt an explicitly diachronic approach, examining each factor's role during three distinct periods: 1960–79, 1980–97 and post-1998.

UAVs' precocious birth and death

Over the course of the 1960s the United States Air Force developed a spectacularly successful drone program. Ultimately, the Air Force procured over 1,000 Model 147 'Lightning Bugs' which conducted a wide-range of missions over North Vietnam, including collecting photographic intelligence, dropping chaff to confuse radars, decoying enemy missiles and even attacking heavily defended ground targets. Precocious to the extreme, drones would not again conduct many of the Lightning Bug missions for decades hence. Political leaders' broad appetite for imagery intelligence to inform their decisions catalyzed the process that ultimately led to the Lighting Bug, but politicians themselves lacked the expertise to determine what platforms to promote. Likewise, although the Air Force managed the Lightning Bug program, its impetus came from that service's rivalry with the CIA and, to a lesser degree, the Army. It was thus the competition of multiple governmental agencies and armed services to fulfill political leaders' desiderata that generated this first drone revolution.

The Air Force traditionally regarded aerial reconnaissance as its bailiwick and had flown photographic reconnaissance missions (as the Army Air Corps) since the First World War.[25] Reconnaissance, however, was never a central Air Force priority and the service generally did no more than modify its existing combat aircraft for this mission. During the Cold War, however, such modified bombers (RB-47 s and RB-57 s) and fighters (RF-101 s) were proving inadequate, being recurrently intercepted and shot-down by Communist fighters. The United States, indeed, lost a total of 23 aircraft and 179 aircrew during peacetime reconnaissance missions from 1946 to 1990.[26] Traditional reconnaissance aircrafts' increasing vulnerability over the course of the 1950s drove civilian policymakers to demand alternative means for obtaining imagery intelligence.

The CIA leapt at the opportunity to fill the resultant gap in national reconnaissance capabilities and worked with Lockheed Aviation's Chief Engineer, Clarence Johnson, to design a specialized reconnaissance aircraft. The resultant aircraft, the U-2, was secretly developed between 1954 and 1956, and could fly higher than any previous aircraft, including 10,000 ft higher than the Air Force's highest-flying aircraft.[27] The U-2's success soon put the CIA in the driver's seat when it came to photographic reconnaissance.

The U-2, however, only guaranteed the CIA preeminence for a short-time. Already in 1957, the Soviet Union began deploying a surface-to-air missile, the SA-2, that would soon put the U-2 at risk. One veteran of the Air Force's drone program described the competition between the CIA and Air Force over aerial reconnaissance as a 'running gun battle' within which 'each organization on a project-by-project basis tried to push the other team out of the picture.'[28] It was, within this context of intense inter-organizational competition that the Air Force embraced drones. Even then, while the Air Force's leadership desperately sought means for reestablishing their position as a photographic intelligence collector, their use of drones to achieve that end was fortuitous.

Indeed, it was Ryan Aeronautical Company – the producer of the Air Force's Firebee target drone – that began without Air Force encouragement to examine modifying its Firebees for reconnaissance purposes in 1959. Ryan's executives then submitted their unsolicited drone proposal to the Air Force in 1960. At this point, the Air Force accorded such a low priority to reconnaissance as a mission and drones as a platform that they lacked a systems program office (SPO) dedicated to either.[29] Ryan's proposal therefore might have languished if not for outside events and inter-organizational pressures.

Ryan's drone proposal, however, was evaluated by Air Force leaders in the immediate aftermath of the Soviets shooting down two reconnaissance aircraft; the May 1960 shoot-down of Gary Power's CIA-operated U-2 and the July 1960 destruction of an Air Force RB-47. The notion of an unmanned and therefore expendable reconnaissance platform proved attractive to Air Force leaders after these two incidents, which had killed four American aircrew and led to one's imprisonment in the Soviet Union. The Air Force therefore provided 200,000 USD for Ryan to pursue work on its reconnaissance drone project, which soon became known as the Model 147 Lightning Bug.[30]

This financially modest contribution hardly signaled the service's full-fledged commitment to drones. Nevertheless, Air Force commanders felt compelled to continue supporting the project lest rival services and agencies once again wrest leadership of photographic reconnaissance from them. Already, Lockheed was studying building an unmanned version of the U-2 aircraft for the CIA and was also designing a stealthy supersonic reconnaissance aircraft – the A-12 – that could outrun enemy missiles.[31] The Army Signal Corps, meanwhile, launched its own reconnaissance drone project – the SD-5 – whose capabilities overlapped with those of the Air Force's proposed Lightning Bug.[32]

It was within this context of inter-organizational competition that the Air Force rushed the Lightning Bug into service. The CIA, indeed, appealed to the Executive Branch to take over the Air Force's project in early-October 1962. The prospect of having their project taken from them prompted the Air Force's leadership to deploy prototype Lightning Bug drones to Florida during the Cuban Missile Crisis and dramatically increase the project's funding in 1963. The Air Force, within this context, formed a well-resourced drone unit that same year, equipped with Lightning Bug drones, modified C-130 transport aircraft to launch them and helicopters to recover them mid-air. The Air Force soon flew these drones as part of the Kennedy Administration's effort to destabilize Cuba – Operation Mongoose – hoping that the drones would provoke Cuban surface-to-air missile units into shooting at them and thereby providing a pretext for military retaliation.[33]

These actions, which ensured the Air Force's continued ownership of the project, coincided with the United States' increasing difficulty gathering intelligence over China. It was consequently by providing intelligence on southern China that the Air Force's Lightning Bug attracted the attention of American political leaders. Cloud cover and forests hindered the ability of the United States' spy satellites and high-altitude reconnaissance aircraft to collect photographic intelligence on southern China. Chinese air defenses – both anti-aircraft guns and fighter aircraft – had meanwhile achieved the proficiency needed to intercept a prohibitively large proportion of low-altitude reconnaissance aircraft. Indeed, they had shot down five RF-101 and P2V manned reconnaissance aircraft, operated by Taiwan's Air Force in partnership with America's Air Force, between 1961 and 1965.[34] Flying U-2s at higher altitudes, likewise, was proving suicidal, as demonstrated by the Chinese destruction of three Taiwanese-operated U-2s in 1964.[35]

The Air Force consequently rushed its Lightning Bug unit to Okinawa in 1964, from whence they flew 15 missions over southern China between August and November.[36] China's government immediately countered by deploying an elite 'division of ace pilots' to counter the UAVs, yet even this unit found intercepting comparatively small, elusive drones challenging.[37] Nonetheless, a Chinese pilot eventually shot down a drone on 11 November 1964.[38] This destruction of a Lightning Bug failed, however, to generate adverse publicity equivalent to that experienced when manned reconnaissance aircraft had been shot down.

Political leaders consequently embraced drones as a politically advantageous means for collecting intelligence and encouraged their development. National Security Adviser Henry Kissinger expressed this view when he opined of the drone operations over southern China, 'The 147 H/T [Lightning Bug] drone is a valuable vehicle for obtaining the type of military tactical intelligence desired. Manned U-2 or SR-71 overflights would be considerably more provocative than drones. Satellite reconnaissance has failed to fill the need [in southern China].'[39] The United States, therefore, continued conducting Lightning Bug missions over South China despite the occasional loss of drones. Altogether, they flew 97 such missions – an average of 19 - per year – between 1964 and 1969.[40]

Even more so than China, Ryan Aeronautical's UAVs came fully into their own during the Vietnam War. As America escalated its involvement in that country, it found itself facing formidable North Vietnamese air defenses. Indeed, according to American intelligence, North Vietnam's arsenal of anti-aircraft guns rose meteorically from 943 anti-aircraft guns of all calibers in 1965 to 6000–7000 large-caliber guns in 1966, all above 20 mm in caliber.[41] Likewise, the North Vietnamese began rapidly introducing SA-2 surface-to-air missiles into their arsenal in 1965: the same system that had proven so deadly against the U-2 in other parts of the world.

It was in this increasingly deadly environment that Lightning Bug began to be truly valuable. Although the Lightning Bug project began modestly as an effort to contest the CIA's expanding role in aerial reconnaissance, the Vietnam War highlighted the drone's potential to the Air Force's combat arms, leading them to promote its continuing development over the war's course. As early as late-1964, Lightning Bugs were conducting reconnaissance sorties out of Bien Hoa against North Vietnamese targets.[42] As North Vietnamese airspace became increasingly perilous, unmanned aircraft became attractive both to politicians worried about the political embarrassment of captured air crew and to the Air Force's reconnaissance community, which was increasingly conscious of their vulnerability.[43]

The Air Force consequently developed new Lightning Bug variants throughout the war, including models for destroying enemy air defenses, dispensing chaff, collecting signals intelligence and decoying enemy surface-to-air missiles. By the war's end, 1,016 Lightning Bugs flew a total of 3,435 sorties with a mission success rate of 84 per cent.[44] In achieving this objective, the Air Force created four drone units and fielded 25 different drone variants. While this represented a fraction of total reconnaissance sorties, it still was the first large-scale use of UAVs in warfare.

The United States' withdrawal from the Vietnam War in 1973 posed questions about drones' future. Drones' successes led many to conclude that now was the time to double-down on unmanned aircraft and envisage their increasing use in front-line combat roles.[45] Drones' combat record in Vietnam, indeed, drove a mushrooming of drone proposals such that 35 distinct Air Force drone programs were underway by 1973, ranging from paper design studies to further modifications of Lightning Bugs. Key amongst these were the BGM-34 C project, which aimed to develop an improved Lightning Bug unmanned combat air vehicle (UCAV) capable of delivering a wide-range of weaponry, and the Compass Cope project, which sought to produce a high altitude, long-endurance drone, equivalent to today's Global Hawk.[46]

Despite the Air Force's recent military experience and its range of ongoing drone projects, the inter-organizational competition that had spurred drones' development in the early 1960s no longer existed. The Army was first to withdraw from the field, cancelling its SD-5 in November 1962. Much more consequentially, however, was the CIA's subsequent withdrawal from the airborne photo-intelligence role. The CIA's own drone – code-named the D-21 – had failed and faced cancellation in 1971. This project's objective of building a stealthy supersonic drone, based on A-12/SR-71 technology, proved beyond what contemporary technology could achieve, leading President Richard Nixon to order its cancellation after flying five failed missions over China.[47] A subsequent Presidential Order, issued in 1976, transferred all of the CIA's aviation assets – manned and unmanned – to the Air Force.

The Air Force, thus, enjoyed a monopoly on drone and aerial photo-reconnaissance capabilities in the Vietnam War's immediate aftermath. With no other agency poised to challenge the Air Force for the mission, that service's leadership could afford to neglect drones. They consequently deactivated three of their four drone units between 1973 and 1976. Then, in 1977, the Air Force cancelled its two premier drone projects – the BGM-34 C and Compass Call – when they experienced cost overruns and reliability problems. Although the technical challenges these programs faced were real, they were arguably not more so than those encountered by other sophisticated weapons programs.

However, in the absence of either a pro-drone community within the Air Force or inter-organizational competition forcing the Air Force to stay the course, the Air Force's leadership seized upon hiccups in both drones' development to cancel them. Finally, then, in 1979 the Air Force disbanded its last drone unit. As a result, without the original incentive structure that had driven drones' development and with no meaningful competition from other organizations, UAV development was allowed to wither on the vine under a brief Air Force monopoly.

Competition and drones' redux

Since the Air Force's post-Vietnam monopoly on UAVs precipitated the cancellation of drone development projects thereafter, the revitalization of

United States military drone capabilities logically occurred once rival services and agencies again invested in unmanned aviation. Indeed, it was first the Navy, followed by the Marine Corps, and then the Army, which acquired reconnaissance drones. These developments rekindled the CIA's interest in drone project as well, leading to the program that ultimately evolved into the Predator drone. Only once it felt that its status as America's premier vector of air power was threatened by other services' drone investments did the Air Force follow suit. While inter-organizational competition drove this renaissance in American unmanned aviation, political leaders spurred this process along by setting clear strategic objectives and then encouraging multiple agencies and services to compete to fulfil them.

Even as the Air Force divested itself of its drones in the 1970s, Israel's armed forces built-up a UAV capability that can be seen as the heir to the Air Force's 1960s drone program. The Soviet Union's deployment of new surface-to-air missiles to Egypt and Syria in 1969 led to the destruction of an increasing number of Israeli reconnaissance aircraft. Having heard of the Lightning Bugs' success in Vietnam, Israeli Air Force Chief of Staff Mordechai Hod contacted Ryan Aeronautical.[48] Ryan, in turn, developed a variant of the Lightning Bug – the Model 124I – to meet Israel's particular requirements, which involved carrying different cameras and launching the drone from the ground, rather than aircraft.[49] Ryan delivered its first Model 124I to Israel in 1971, which proved invaluable against Arab air defenses during the 1973 Arab-Israeli War. Israeli military commanders appreciated their Lightning Bug derivates to such a degree that they purchased much of the United States Air Force's stockpile when the latter ultimately sold them for scrap.[50]

However, Israel's armed forces had to rely on indigenous firms for their follow-on drone projects since the United States had left this domain. Two Israeli firms, IAI and Tadiran, began developing inexpensive UAVs following the 1973 Arab-Israeli War. One reason Arab air defenses had proven formidable during that war was that their SA-6 missiles were mounted on armored vehicles and regularly shifted locations. Israel's Air Force had trouble tracking and therefore destroying these mobile missiles. The IAI Scout and Tadiran Mastiff UAVs solved this problem after they entered service in 1978 by enabling personnel in Israel to remotely pilot the small UAVs for six-hour missions to scan terrain for mobile missiles. The UAVs' cameras instantaneously transmitted images back to base, enabling planners to plan strikes against fleeting targets.[51]

Although Israel's Mastiffs and Scouts were less costly and sophisticated than the American-made Lighting Bugs – featuring propellers rather than jet engines – they provided the cost-effective edge that Israel needed when it next went to war. Israeli firms clandestinely developed these technologies, as well as new electronic jamming pods, in under five years. Israel put these SAM-suppression adaptations to the test during the 1982 Lebanon War.

Syrian President al-Asad deployed 19 SA-6 missile batteries – more SA-6 s than Syria had possessed in 1973 – to Lebanon's Beka'a Valley in April 1981 to deter Israel from intervening.[52]

Israel's UAVs, however, emboldened Defense Minister Ariel Sharon and Chief-of-Staff Raphael Eiten to argue in 1982 that Israel could win a swift victory, at little cost, by first destroying Syria's air defenses and then pummeling hostile ground forces. Israeli UAVs subsequently ferreted out Syrian SAM batteries' positions in the months before Israel's invasion. Israel's Air Force then attacked on 9 June, destroying 17 of Syria's 19 SAM batteries, while suffering no losses of their own. This offensive eradicated Syria's SAM network in a single stroke, rendering Syrian ground forces defenseless against Israeli air strikes.[53]

Israel's facile success with UAVs spurred emulation by all the United States' armed services, with the notable exception of the Air Force. The Navy led the way in embracing Israeli-style drones after having failed in its attempts at 'gunboat diplomacy' in Lebanon in 1983. After America's government had intervened in Lebanon, the Navy sought to use its large warships – battleships and aircraft carriers – to coerce the Syrian Army and various Lebanese militias into cooperating. At a technical level, however, American sea power proved impotent. Battleship cannons could theoretically bombard targets at a distance of 39 km, yet battleships had no means for observing where their shells landed at that distance, which meant that they often missed their objective. When the Navy used jets to attack Syrian artillery, the result was even worse, with the Syrians shooting down three American aircraft in a single day.[54]

The contrast between the Israelis' facile destruction of Syria's air defenses and the Navy's humiliation at those same Syrians' hands drove Navy Secretary John Lehman and Marine Corps General Al Gray to successfully lobby in 1984 for the purchase of small numbers of Mastiff UAVs for testing. Experimentation with these Mastiffs led the Navy and Marine Corps to commission Israel's Tadiran firm to develop an improved Mastiff – the RQ-2 Pioneer – for them.[55] Lehman – a politically appointed civilian – proved the most enthusiastic advocate of the Navy's drone program, encouraging the Navy's surface warfare community's interest in drones and combating naval aviators' opposition to them. Tadiran and its American joint-venture partner, AAI Corporation, delivered the first eight Pioneers to the Navy in 1986, whence they were deployed aboard battleships. The Marine Corps subsequently received Pioneers in 1987 and the Army acquired them beginning in 1990.[56]

Although the Navy, Marine and Army inventories of Pioneers were comparatively modest, all three services employed their UAVs for the 1991 Gulf War. During that conflict, Pioneer UAVs scouted for the United States' battleships, which spectacularly bombarded Kuwait's coast with their 16-inch guns, and flew reconnaissance missions in support of advancing Marine and Army units. Altogether, the Marines' Pioneers flew 323 missions, the Navy's flew 83 and the Army's 46.[57] Beyond their military utility, UAVs captured the public imagination, which witnessed their being launched from battleships live on television. News reports, further, highlighted an unprecedented incident, when a group of Iraqi soldiers attempted to surrender to a Pioneer; marking the first known instance of human soldiers capitulating to an unmanned aircraft.[58]

Post-war evaluations emphasized how valuable UAVs were. One Marine Corps intelligence officer, for example, told Congressional staffers that his division could have used three times as many drones as they possessed.[59] General Walter Boomer, commander of Marine forces during the Gulf War, further praised the Pioneer as that War's 'single most valuable intelligence collector.'[60] The Pioneers' success in the Gulf War drove the Army to seek a more robust reconnaissance UAV of its own. The resultant UAV-SR (Short Range) competition of 1991–1993, in turn, led to the Army's selection of a drone – the Hunter – designed by Israel Aerospace Industries.[61] The Army purchased 56 Hunters, in turn, which began entering service in 1995.

Even as the Navy, Marines and Army cultivated modest drone capabilities through the purchase of modified Israeli systems, other United States government agencies promoted drones as well. The Defense Advanced Research Projects Agency (DARPA), a Defense Department agency dedicated to cultivating advanced technologies for eventual use by the armed forces, sponsored a succession of reconnaissance drone projects in the 1980s and 1990s. DARPA funded the development of the massive Condor UAV by the Boeing Corporation in the 1980s. This large (9,000 kg) high-altitude (67,000 ft) drone was designed to offer an inexpensive alternative to spy satellites.[62] DARPA later launched its advanced airborne reconnaissance program in 1993, which funded Lockheed Martin's development of the stealthy Darkstar drone and Boeing's high altitude Global Hawk.[63] Despite DARPA's efforts to fund advanced drones, the armed services long hesitated to adopt DARPA's programs as their own and fund their development into deployable military systems.

One of the DARPA projects, however, would ultimately evolve into the Predator project, which became the single most momentous drone project of the period. In 1984, DARPA contracted Abraham Karem, an Israeli engineer who had emigrated to the United States, with developing an experimental UAV called Amber. Amber, in principle, would be a propeller-driven reconnaissance UAV that could also attack targets of opportunity by jettisoning its wings and arming a warhead in its nose.[64] The 5 USD million DARPA grant, however, proved insufficient to fund Amber's development or support Karem's start-up company. Karem, consequently, sought to develop a less sophisticated observation drone, without Amber's kamikaze capability, to sell to foreign clients. This drone, dubbed the Gnat-750, failed to save Karem's firm, which was bought by Hughes Aviation, and then was sold again to General Atomics.

The wars that broke out in the former Yugoslavia in the early 1990s, however, won Karem's Gnat-750 attention from the CIA. Ubiquitous cloud cover and mountainous terrain meant that the United States' surveillance satellites frequently failed to furnish adequate intelligence on developments in the Balkans. Serbian air defenses, meanwhile, dissuaded the United States from flying low-level manned reconnaissance missions. President Bill Clinton personally favored the United States' playing a larger role in resolving these conflicts and his cabinet therefore prioritized obtaining high-quality photographic intelligence in the Balkans. CIA Director James Woolsey, within this context, directed his agency's technical services to partner with General Atomics in 1993 to expeditiously deploy Gnat-750s for reconnaissance. By February 1994, a mixed team of CIA personnel and General Atomics employees operated Gnat-750 s over Bosnia out of a base in Albania.[65]

Faced-paced testing of the Gnat-750, however, revealed significant scope for improving the UAV. DARPA, consequently, founded the Gnat-750's redesign, with a larger airframe and a satellite communications link, into the UAV now known as the Predator. Approved under the Defense Department's Advanced Concept Technology Demonstration (ACTD) rules, which allowed for a faster acquisition strategy, the Predator was fielded with remarkable speed.[66] After an initial demonstration in 1994, the aircraft was flying reconnaissance missions in July 1995.[67] Although limited by adverse weather, the Predator proved superior to manned aircraft when it came to gathering intelligence over large areas for extended periods of time.

As the Predator program progressed, however, a traditional military service – the Army – began to covet it. The Army had long toyed with unmanned aviation as a means of compensating for what it perceived to be the low priority that the Air Force attached to supporting ground forces. Congress, however, cancelled the Army's principle UAV program to date – the Aquila drone – in 1987, after eight years of development and amidst claims that the program had been poorly managed.[68] The Aquila's objective of providing both an inexpensive reconnaissance platform and a system capable of directing laser-guided artillery shells proved more than contemporary technology could accomplish.

After the Aquila's cancellation, the Army followed the Navy's lead in purchasing Pioneers and then purchased Israeli-designed Hunters as well. The Army's leadership, therefore, naturally proposed the services of their Military Intelligence Battalion (Low Intensity) for testing the Predator.[69] It was thus under the Army that the Predator made its operational debut in 1995 – first in the multi-national Roving Sands military exercise and then in providing intelligence over Bosnia.[70]

It was only at this stage, once the Predator's viability had been proven over the Balkans and a rival military service had expressed interest in it, that the Air Force sought to appropriate the program. The Predator's development until this point benefitted from support by DARPA, the CIA and then the Army. The Navy and Marine Corps, meanwhile, demonstrated their curiosity about the system by sending one pilot apiece to observe the system's debut in the Balkans. The Air Force, indeed, was conspicuous for being the last armed service to express interest in the Predator program.

At the point when the system was poised to enter full-scale production, however, the specter of a rival organization developing a sizeable airborne reconnaissance capability drove the Air Force's leadership to seek to appropriate the project. Air Force Chief of Staff, General Ronald Fogleman, consequently ordered the creation of the 11th Reconnaissance Squadron in July 1995: the first operational Air Force UAV squadron since the last one's disbandment in 1979.[71] Fogelman, however, recognized that the Air Force needed to appear even more bullish on drones or else other services and agencies would encroach in this domain on the Air Force's status as America's premier employer of airborne systems.

He therefore overruled the 11th Reconnaissance Squadron's first commander by advocating for the Predator's rapid acquisition even after the latter had cautioned that the Predator was still too unreliable.[72] He also articulated an ambitious Air Force strategy for UAVs, wherein the Predator would provide an interim capability until the Air Force could procure the stealthy DarkStar and high altitude Global Hawk, whose development DARPA had initiated. To this end, Fogelman established an Air Force Systems Program Office (SPO) in late-1995 tasked with taking over these two UAV development programs from DARPA.[73] The Air Force's leadership subsequently argued to political leaders that they should assign the Air Force ownership of the burgeoning Predator program and, more broadly, recognize it as the United States' lead service for drones.

Fogelman's politically savvy gambit succeeded beyond his own expectations in 1996 and 1997. Defense Secretary William Perry decided in April 1996 to 'give' the Air Force the Predator program, permitting the Army to continue operating the drones over the Balkans only until the Air Force could train its own personnel to take over.[74] The House Intelligence Committee, meanwhile, ruled in 1997 that the CIA should hand over all of its functions related to the Predator to the Air Force.[75] DARPA, meanwhile, prepared to transfer responsibility for the DarkStar and Global Hawk to the Air Force in the late-1990s.[76]

The Air Force thus appeared poised in the late-1990s to reestablish both a robust UAV capability and the monopoly on drones that it had enjoyed in the 1970s. This development came on the heels of a decade of heady drone development that had been fueled by inter-organizational competition. Although political leaders promoted this process at critical junctures, such as when Naval Secretary Lehman pushed to acquire Israeli drones or when President Clinton prioritized obtaining intelligence in the Balkans, it was the competition between the Navy, Marines, Army, DARPA and the CIA to meeting political leaders' desiderata that generated innovation. The Air Force, however, long remained the most reticent service when it came to UAVs. Indeed, only the real prospect of two rival organizations – the Army and CIA – developing the Predator into a significant airborne asset finally drove the Air Force to embrace drones and seek leadership over their subsequent development.

Consolidating the innovation

Questions might legitimately have been asked as to how the United States' drone projects would evolve after the febrile period of drone development in the 1980s and 1990s. The Air Force, under Fogleman's command, had belatedly, yet enthusiastically embraced drones. Espousing a more ambitious vision of drones than rival organizations, the Air Force had provisionally won control over the United States' premier drone projects – Predator, Global Hawk and DarkStar – in 1996–97. Would the Air Force, however, use its newly reestablished preeminence to aggressively develop drones or, rather, scale-back its drone efforts as Air Force leaders had done in the late-1970s? On this occasion, inter-organizational rivalry obliged the Air Force to continue investing in UAVs. Civilian policymakers and rival organizations alike contested the Air Force's pretensions towards a monopoly on drones and thereby obliged it to continue pursuing the innovation.

No sooner had the Air Force won control of the Predator program in 1996 than the Army began manifesting its discontent with the arrangement. Air Force pilots assigned to Predator indeed reportedly showed contempt for the unmanned vehicle during the handover meeting between the Air Force officers who were scheduled to take over Predator and the Army officers who had been operating it.[77] This led the Predator's Program Officer to question, 'Maybe Fogleman really wanted Predator, but maybe Air Force leaders just wanted to keep another service from having it.'[78] Army units engaged in peacekeeping in Bosnia then complained in 1997 that the Air Force's Predators were not adequately supporting them.[79]

Claims that the Air Force was misusing Predator prompted Air Force leaders to redouble their commitment to it. Fogleman, to this end, selected a rising star within the Air Force, Colonel James Clark, to take over the 11th Reconnaissance Squadron in 1997.[80] Improving Predator subsequently became an Air Force priority and the service endeavored to prove its competence. During the 1999 Kosovo War, Air Force Predators demonstrated a greatly improved ability to broadcast images directly to NATO and American headquarters by satellite. Air Force, technicians, meanwhile, operated their Predators aggressively; collecting imagery below the Balkans' cloud cover that other reconnaissance assets – satellites and manned aircraft – could not.

Summarizing the Predators' accomplishments over Kosovo, RAND Corporation analyst Benjamin Lambeth enthused that,

> [F]or the first time in American combat experience, UAVs offered commanders and planners the frequent advantage of real-time video imagery without any accompanying danger of aircrew losses … . Several UAVs were lost when commanders requested closer looks, forcing the drones to descend into the lethal envelopes of Serb AAA and man-portable air defense systems (MANPADS). These losses did not evoke great concern, however, since the UAVs were intentionally sent out on missions that were known ahead of time to be especially risky.[81]

Operated in this way, the Air Force lost four Predators to Serbian air defenses; a figure which was high compared to the losses of NATO's manned aircraft, of which two were lost during war, yet still considered acceptable.[82]

The Air Force also frenetically developed a laser designator to enable its Predators to guide laser-guided bombs dropped by aircraft flying above the cloud cover onto targets that the aircraft themselves could not see. So expeditiously was this laser-equipped Predator developed that a prototype conducted a mission over Kosovo on the war's penultimate day. The Air Force's commander during the Kosovo War, General John Jumper, soon advocated expanding the Predator program. Jumper's frustration that Predators frequently spotted enemy targets, yet no friendly aircraft was available to strike them, led him to champion arming Predators in March 2000.[83]

From his new position as head of Air Combat Command, Jumper pushed Air Force engineers to weaponize Predator. These latter, in turn, identified the Hellfire anti-tank missile as a suitable munition and began modifying both the drone and the missile for use together. Anticipating the Air Force's arming of Predators, General Atomics began to develop a technology demonstrator of a scaled-up Predator in 1999 – later known as Reaper – designed specifically to carry larger payloads and more weaponry.[84] That same year of 1999, meanwhile, the Air Force partnered with DARPA and NASA to sponsor a stealthy, sophisticated Unmanned Combat Aerial Vehicle (UCAV) project. This project aimed to explore the next generation of armed drones by contracting with Boeing to develop the X-45.[85]

Other organizations, however, took umbrage at the Air Force's pretensions to a monopoly on drones. The Navy's leadership, for its part, responded by seeking to establish a commanding position over the next generation of UAV technology and thereby preclude the Air Force from perpetuating the stranglehold it had established over the Predator project. To this end, they contracted Lockheed Martin in 1997 to analyze the prospects for a stealthy Unmanned Combat Aerial Vehicle (UCAV), dubbed the Uninhabited Naval Strike Aircraft (UNSA).[86] The Navy subsequently partnered with DARPA in

2000 to sponsor a stealthy UCAV project equivalent to the X-45. This project, termed the X-47, aimed to test technologies for operating sophisticated UCAVs from aircraft carriers.[87]

The Navy's rivalry with the Air Force over the next generation of drones coincided with CIA's efforts to break the Air Force's monopoly over Predator. Agents at the CIA's Counterterrorism Center sought means for collecting intelligence on Osama bin Laden and his compatriots in Afghanistan after American cruise missiles had failed to inflict meaningful damage on al-Qaeda in 1998. Predator's effectiveness in Kosovo in 1999 prompted the Counterterrorism Center's leaders to lobby for the use of Predator over Afghanistan in 2000 to pinpoint bin Laden's whereabouts, a mission that the drone successfully accomplished in September 2000.[88] The unarmed Predator's accomplishment at collecting imagery of bin Laden, however, prompted some to argue that the CIA needed access to armed Predators, with which they could assassinate figures such as bin Laden if the opportunity arose. The debate about whether the CIA should acquire armed Predators thus raged throughout early-2001, with CIA Director George Tennet opposing the move and National Security Council's Counter-terrorism Coordinator, Richard Clarke, and the CIA's Counterterrorism Center favoring it.[89]

The Air Force's preeminence in the domain of drones was thus already a contested reality when al Qaeda operatives attacked the United States on 11 September 2001. These attacks and the United States' subsequent invasions of Iraq and Afghanistan, however, transformed the inter-organizational struggle over UAVs by dramatically increasing the resources involved. For example, whereas the United States armed forces possessed a total of 82 drones in 2001, they would own approximately 8,000 by 2010.[90] Civilian policymakers were particularly bullish on drones in the so-called 'War on Terror' because of the reduced political costs for operating them over neutral countries.

Political leaders consequently pushed the armed forces to prioritize these systems. The Senate Armed Services Committee in 2001 even set as a goal that one-third of the military's deep strike aircraft consist of unmanned UCAVs by 2010.[91] The Air Force's leadership, within this context, showed much less enthusiasm for drone than their political leaders. For example, in fiscal year 2003, the Air Force requested a mere 23 USD million for seven new Predators only to have Congress instead appropriate 131 USD million for the purchase twenty-nine.[92] Even more worryingly from drone proponents' perspective, the Air Force failed to promote any drone programs as follow-ons to the Predator and Global Hawk projects that it had taken over in the late-1990s. Air Force leaders, indeed, cancelled the futuristic X-45 project, which had evolved into the J-UCAS program, in 2006 after escalating the performance requirements that it insisted the UCAV meet.[93] They also resisted reallocating money from their U-2 program to accelerate Global Hawk's development.[94]

Dissatisfaction with the Air Force's approach led political leaders to increase their support for rival organizations. The CIA was first to benefit from this dynamic. President George W. Bush expeditiously approved the CIA's acquiring an extensive drone fleet and using those drones for assassinations both over countries where the United States was officially conducting combat operations as well as others where it was not.[95] On 8 October 2001 the CIA consequently conducted its first drone strike – a failed attempt on Afghan leader Mullah Omar – which was followed in mid-November by the successful strike on al Qaeda's third highest-ranking official, Mohammed Atef.[96] The CIA's involvement with drones subsequently grew such that the Agency, by 2014, owned 80 Predators.[97]

Other armed services also benefitted from civilian leaders' dissatisfaction with the Air Force. The Army, for example, obtained Congressional approval for producing a new tactical UAV, the RQ-7 Shadow, in 2003 and a smaller man-portable UAV, the RQ-11 Raven, in 2005.[98] More ambitiously, the Army invited defense firms to submit proposals in 2002 for a larger Extended-Range Multi-Purpose UAV for surveillance purposes. Army leaders then selected a drone, named Warrior and then Grey Eagle, which was in actuality a slightly larger version of Predator. The Navy, meanwhile, launched an ambitious Broad Area Maritime Surveillance (BAMS) requirement in 2003, which envisioned the use of sophisticated, large UAVs to monitor the world's oceans. The first phase of BAMS envisaged the Navy acquiring Global Hawk UAVs before then developing a more specialized platform.[99] With the Navy acquiring Global Hawks, and the Army and CIA Predators, the Air Force's primacy within the drone arena was fading fast.

Restoring the Air Force's monopoly over the use of UAVs became one of the service's priorities in 2006. To do this, the Air Force's leadership argued that their service should become the executive agency of all UAVs flying above 3500 ft and that the next wave of long endurance UAVs should be developed and managed under their aegis.[100] The Air Force's leadership argued to this effect that theirs' was the most experienced service at handling UAVs and also possessed superior expertise in the related domains of cyberspace and satellite communications.[101] Understandably, the Navy's and Army's representatives bristled at the Air Force's attempt to claim ownership over their programs, igniting a public debate over who should control UAVs.[102]

Defense Secretary Robert Gates intervened at this point to decisively thwart the Air Force's bid for control. Gates, indeed, provocatively criticized the Air Force's pilot-centric culture and encouraged them to follow the Army's example in adapting to new innovations in a 2008 speech at Maxwell Air Force Base.[103] He also announced the creation of a joint UAV task force that would hasten the fielding of UAVs for the Iraq and Afghanistan wars.[104] Gates's advocacy for the Army extended to supporting them in developing a self-sufficient surveillance unit known as Task Force ODIN, a move that was in direct conflict with the Air Force's monopolization strategy.[105] Finally, Gates's most dramatic move against the Air Force came in his simultaneous firing of both the Air Force's Secretary and Chief of Staff in June 2008. While that move came as a response to a range of scandals, it also symbolized Gates's adversarial relationship with the Air Force's leadership.[106]

With no single service enjoying a monopoly, they all have continued competing for the roles and missions associated with drones. For example, although neither the Air Force nor the Navy have procured a stealthy UCAV capable of substituting for manned combat aircraft, both services have advanced a succession of related projects. The Navy, to this end, is developing a stealth flying wing – the MQ-25 – to refuel its carrier-based combat aircraft mid-flight.[107] The Air Force, for its part, is developing both so-called 'loyal wingman' drones to support manned combat aircraft flying alongside and long-range bomber drones.[108] The Army and CIA, meanwhile, are innovating in how they employ drones. Although the CIA's drone campaign remains cloaked in secrecy, the Agency is reportedly employing new techniques to combine and analyze data from multiple sources to identify high-value insurgents to strike with their drones. The Army, for its part, is seeking to equip each of its active-duty divisions with a company of Hellfire-armed Grey Eagles.

Conclusion

Inter-organizational rivalry has been critical to driving the United States' development of drones. All four of America's armed services have played roles in this process, as did the CIA and DARPA. Indeed, drone development was rapid whenever two or more organizations competed to field operational UAVs. Drone development was contrarily sluggish to non-existent whenever a single organization exercised a monopoly. This finding that inter-organizational competition drove drone innovation in the United States does not suggest that civilian leaders and military bureaucracies played negligible roles. Rather, both fulfilled necessary functions, but were in themselves insufficient to generate innovation in the absence of inter-organizational competition.

The organization that objectively played the principle role in introducing sophisticated drones into service was, however, the Air Force. The Lightning Bug, Predator, Reaper and Global Hawk all, within this context, achieved fully operational status under the Air Force's aegis. The apparent paradox of inter-organizational competition driving innovation, yet the Air Force's playing role in the fielding of more sophisticated drones, demonstrates how competition as a mechanism provides an impetus for innovation. As the only military service dedicated to the pursuit and employment of air power, no other service can equal the Air Force in terms of the size of its procurement bureaucracy dedicated to aircraft or the depth of its relationship with aerospace firms. One would therefore expect the Air Force under *ceteris paribus* conditions to be the most competent service when it comes to managing drone projects.

The Air Force's domination by pilots of manned combat aircraft, however, also aggravated the service's biases against drones. In many respects, these motivated biases are subtle, rather than explicit. Thus, rather than outright seeking to cancel drone projects to protect piloted aircraft, Air Force leaders have rather perceived drones' shortcomings more acutely than those of manned aircraft. Consequently, when resources are finite and projects' futures' uncertain, Air Force leaders tend to cancel unmanned systems rather than equally problematic manned aircraft. The Congressional Research Service characterized the Air Force's outlook as being one of relishing drones as a complementary asset to manned aircraft, but rejecting them as a possible substitute for them.[109]

The Air Force's motivated biases have always had considerable scope for expression. Such is the case because the trade offs between drones and manned aircraft are often ambiguous. Unmanned aircraft are often touted for their lower acquisition costs and ability to patrol for extended periods over battlefields. Manned aircraft, however, are often more capable and have lower accident rates. Unmanned aircraft, moreover, often require more significant ground infrastructures, which challenge facile calculations about cost-effectiveness. When future systems are being considered, the issue as to whether a manned or unmanned option will prove superior often hinges on the fate of as yet immature technologies, such as artificial intelligence or data links. Indeed, drones' one unambiguous advantage – the lower political costs of their use because of the absence of pilots who can be captured or killed – is of such a nature that Air Force officers are likely to give it short-shrift in peacetime.

Given this persistent ambiguity about unmanned and manned aircraft's relative merits, is it any wonder that an Air Force led by combat aircraft pilots would systematically opt for manned solutions? Inter-organizational competition alone can drive the Air Force to promote unmanned solutions under such circumstances. Indeed, the prospect of another service or agency encroaching on the Air Force's position as the United States' principle operator of military aircraft through the successful fielding of UAVs can exercise a powerful influence over the Air Force's leadership. It was thus the CIA's competition as a provider of aerial reconnaissance that prompted the Air Force to embrace the Lightning Bug and the prospect of the Army acquiring Predator that led the Air Force to seek control of the project.

Competition between rival organizations was thus central to drone innovations and innovation consequently ceased whenever the Air Force exercised a monopoly. It was civilian leaders' actions, however, that determined the incidence of such 'creative' competition. Innovation, within this context, occurred whenever political leaders set forth clear operational objectives and then encouraged multiple organizations to compete to fulfill them. The priority the Kennedy and Johnson administrations attached to collecting intelligence over China and President Clinton's demand for imagery in Bosnia constitute two cases in point of political leaders indirectly promoting drone innovations by setting strategic goals and then selecting amongst the proposals to achieve them. At times, such as Navy Secretary Lehman's advocacy of Israeli drones or the Senate's setting of an ambitious 'headline goal' for UCAV procurement, civilians tried to promote drones more proactively.

On balance, however, the indirect approach of encouraging inter-organizational competition proved more fruitful to innovation than the direct one of prodding armed services to invest in drones. Civilian policymakers eventually recognized this reality and they consequently exploited inter-organizational competition to oblige armed services to provide them with capabilities they consider essential. The Congressional Research Service articulated this growing consciousness on political leaders' part by observing that, 'if UAS [Unmanned Aerial System] efforts are too centralized [by the Air Force] ... competition and innovation may be repressed.'[110]

This finding – that inter-organizational competition provided the primary impetus for drones' development – is robust even considering drones' rapid development in wartime. At first glance, American drones' astonishing progress during the Vietnam War and, later, during the so-called War on Terror suggests that military organizations develop drones slowly in peacetime, yet rapidly in wartime. In both cases, small peacetime Air Force inventories that included modest numbers of specialized reconnaissance drones burgeoned under the pressure of urgent operational requirements into large diversified drone forces. Nonetheless, this rapid wartime expansion of Air Force drone capabilities was only possible due to the inter-organizational competitions that preceded the outbreak of war. Indeed, the drone models that became emblematic of the two conflicts – the Lightning Bug in Vietnam and the Predator during the War on Terror – were both quintessential products of inter-organizational competition and were already in service when those conflicts began. In other words, while wartime exigencies fueled dramatic growth in unmanned aviation, this was only possible because competition in peacetime generated the systems that made that growth possible.

While drones – the innovation featured in this study – may appear at first glance to be a technology whose development is particularly prone to inter-organizational competition since multiple agencies and organizations possess aviation assets, there are strong reasons for anticipating that a large and growing range of national security capabilities may be subject to this form of innovation. As mentioned in the introduction, paramilitary police forces – ranging from

France's *gendarmerie* to China's People's Armed Police – compete with conventional ground forces for missions such as peacekeeping, counterinsurgency and counter-terrorism. Coast guards and maritime militias, meanwhile, vie with navies for roles in upholding states' maritime boundary claims and patrolling their exclusive economic zones (EEZs). Emerging technologies, such as cyberwar and artificial intelligence, finally, offer new horizons for inter-organizational competition since responsibility for these domains in any give state is generally diffuse across multiple military services and intelligence agencies.

In sum, inter-organizational competition has already been critical to drones' emergence as a major military innovation and is likely to play an even larger role in the future given the evolving technologies and security challenges that states face.

Notes

1. Ailleret, *L'art de la guerre et la technique*, 23.
2. Posen, *The Sources of Military Doctrine*.
3. Ibid., 220–44.
4. Zisk, *Engaging the Enemy*; and Rynning, *Changing Military Doctrine*.
5. Evangelista, *Innovation and the Arms Race*.
6. Rosen, *Winning the Next War*.
7. Corum, *The Roots of Blitzkrieg*.
8. Evans and Peattie, *Kaigun: Strategy, Tactics, and Technology*.
9. Kuehn, *Agents of Innovation*.
10. Gundmundsson, *Stormtroop Tactics*.
11. Doubler, *Closing with the Enemy*.
12. Katzenbach Jr., "The Horse Cavalry."
13. Ibid., 149.
14. Builder, *The Masks of War*.
15. Kier, *Imagining War*; and Johnson, *Fast Tanks and Heavy Bombers*.
16. Humphreys, *The Way of the Heavenly Sword*; and Cameron, *American Samurai*.
17. Coulam, *Illusions of Choice*.
18. Armacost, *The Politics of Weapons Innovation*.
19. Coté, *The Politics of Innovative Military Doctrine*.
20. Ibid., 79.
21. Ibid., 77.
22. Richelson, *The Wizards of Langley*.
23. Tanguy, *Missions Extrêmes*, 11–27.
24. Ibid., 31–79.
25. Newcome, *Unmanned Aviation*, 35–38.
26. Ibid., 71.
27. Peebles, *Shadow Flights*, 96–132.
28. Miller, *The 99th Strategic Reconnaissance Squadron*, 38.
29. Ibid., 25–26.
30. Ibid., 26.
31. Rich and Janos, *Skunk Works*, 262.
32. CIA, "SD-5 Drone Characteristics and Capabilities."

33. Miller, *The 99th Strategic Reconnaissance Squadron*, 61–70.
34. Xiaoping, *The PLA Air Force*, 54–70.
35. Ehrhard, *Air Force UAV's*, 9.
36. NRO, Annual Report on Activities of the National Reconnaissance Program.
37. Xiaoping, *The PLA Air Force*, 71–72.
38. Ibid., 73.
39. Kissinger, Air Reconnaissance of South China.
40. Xiaoping, *The PLA Air Force*, 71.
41. Emerson, *Air War Over North Vietnam*, 49–50.
42. Wagner, *Lightning Bugs*, 70.
43. Ibid., 97–98.
44. Newcome, *Unmanned Aviation*, 86.
45. Barkan, "The Robot Air Force is About to Take Off," 281.
46. Wagner and Sloan, *Fireflies and Other UAVs*, 117–18; and Ehrhard, *Air Force UAV's*, 32–34.
47. Rich and Janos, *Skunk Works*, 262–70.
48. Wagner and Sloan, *Fireflies and Other UAVs*, 50.
49. Ibid., 51–54.
50. Ibid., 63–64.
51. Cohen, *Israel's Best Defense*, 429.
52. Evron, *War and Intervention in Lebanon*, 60–104.
53. Clary, *The Bekaa Valley – A Case Study*, 9–11.
54. Halloran, "US Buys Israeli Pilotless Planes."
55. Yenne, *Attack of the Drones*, 37.
56. Ibid., 53.
57. Ibid., 54.
58. Carr and Keane, "A Brief History of Early Unmanned Aircraft," 569.
59. Refer to: "Intelligence Successes and Failures in Operations Desert Shield/Storm." Report of the Oversight and Investigations Subcommittee, Committee on Armed Services (U.S. House of Representatives, 1993), 9. Available at: https://apps.dtic.mil/dtic/tr/fulltext/u2/a338886.pdf.
60. Yenne, *Attack of the Drones*, 54.
61. Ibid., 49–51.
62. Ibid., 47–49.
63. Ibid., 71–76.
64. Fuller, *See it, Shoot it*, 105.
65. Ibid., 112–13.
66. Jones, *Unmanned Aerial Vehicles*, 43.
67. Ibid., 32.
68. GAO, *Aquila Remotely Piloted Vehicle.*
69. Whittle, *Predator: The Secret Origins of the Drone Revolution*, 91.
70. Ibid., 99–101.
71. Ehrhard, *Unmanned Aerial Vehicles*, 541.
72. Ibid., 543.
73. Testimony of Charles Heber (DARPA), Senate Armed Services Committee, 9 April 1997. Available online at: https://www.darpa.mil/attachments/TestimonyArchived(April%209%201997).pdf.
74. Whittle, *Predator: The Secret Origins of the Drone Revolution*, 111.
75. Fuller, *See it, Shoot it*, 114.
76. Testimony of Charles Heber.

77. Whittle, *Predator: The Secret Origins of the Drone Revolution*, 113.
78. Ibid., 112.
79. Ibid., 113.
80. Ibid.
81. Benjamin Lambeth, *NATO's Air War for Kosovo*, 94.
82. Ibid., 97.
83. Fuller, *See it, Shoot it*, 116.
84. Merlin, *Ikhana: Unmanned Aircraft System*, 4–5.
85. Harkins, *X-45: Uninhabited Combat Air Vehicle*, 19–21.
86. Ibid., 10.
87. Ibid., 75–83.
88. Whittle, *Predator: The Secret Origins of the Drone Revolution*, 148–57.
89. Ibid., 182–231.
90. Ibid., 299.
91. Bone and Bolkcom, *Unmanned Aerial Vehicles*, 8.
92. GAO, *Improved Strategic Planning*, 2.
93. Harkins, *X-45: Uninhabited Combat Air Vehicle*, 79–83.
94. Bone and Bolkcom, *Unmanned Aerial Vehicles*, 35.
95. Mayer, "The Predator War."
96. Whittle, *Predator: The Secret Origins of the Drone Revolution*, 247–93.
97. "Just How Many Predator Drones."
98. GAO, *Improved Strategic Planning*; and Bone and Bolkcom, *Unmanned Aerial Vehicles*, 30–31.
99. Bone and Bolkcom, *Unmanned Aerial Vehicles*, 41–43.
100. Weiner, "Evolution in the Post-Cold War Air Force," 113.
101. See House Armed Services Committee, Hearing on Budget Request on Unmanned Aerial Vehicles (UAV) And Intelligence, Surveillance, and Reconnaissance (ISR) Capabilities (19 April 2007), 25. Available online at: https://fas.org/irp/congress/2007_hr/uav.pdf.
102. Ibid., 32–35; and Sevastopulo, "US Military in Dogfight Over Drones."
103. Gates, *Secretary Gates Remarks at Maxwell-Gunter Air Force Base*.
104. Miles, "Gates Forms Task Force to Promote Intelligence."
105. Shanker, "At Odds with Air Force."
106. Barnes and Spiegel, "Air Force's Top Leaders are Ousted."
107. LaGrone, "U.S. Carriers Need New Lethal Unmanned Aircraft."
108. Sherman, "Pentagon Sets Plan for New Bomber"; and Stevenson, "'Loyal Wingman' Part of the Future of Air Combat."
109. Bone and Bolkcom, *Unmanned Aerial Vehicles*, 8.
110. Gertler, *U.S. Unmanned Aerial Systems*, 10.

Acknowledgements

I would like to thank the many individuals who helped bring this research to its completion. Six extremely talented research assistants—Gemma Walls, Joseph Duncan, Oona Singh, Carine Ghannoum, Eilidh Urquhart, and Rory Langan—provided invaluable help when it came to assessing the diffusion of drone technology beyond the United States. Matthew Valla, meanwhile, contributed immensely to my understanding of drone technologies and spent much time debating with me the relative weight of inter-service versus inter-agency rivalry. Ash Rossiter, Jordan Becker, Michael

Hunzeker and Olivier Schmidt have all contributed to my evolving thinking on military innovation more broadly, with Ash deserving special acknowledgements for organizing this special issue. While all of these individuals contributed to my thinking, any errors are my own.

Disclosure statement

There are no conflicts of interest.

Bibliography

Ailleret, Charles. *L'art de la Guerre et la Technique*. Limoges: Charles-Lavauselle, 1950.

Armacost, Michael. *The Politics of Weapons Innovation: The Thor-Jupiter Controversy*. New York: Columbia UP, 1969.

Barkan, Robert. 1972. "The Robot Air Force Is about to Take Off." *The New Scientist*, August.

Barnes, Julian, and Peter Spiegel. 2008. "Air Force's Top Leaders are Ousted." *Los Angeles Times*, June 06.

Bone, Elizabeth, and Christopher Bolkcom. *Unmanned Aerial Vehicles: Background and Issues for Congress*. Washington, DC: Congressional Research Service, 2003.

Builder, Carl H. *The Masks of War: American Military Styles in Strategy and Analysis*. Baltimore, MD: Johns Hopkins University Press, 1989.

Cameron, Craig. *American Samurai: Myth, Imagination, and the Conduct of Battle in the First Marine Division, 1941–1951*. Cambridge: Cambridge UP, 1994.

Carr, Stephen, and John Keane. "A Brief History of Early Unmanned Aircraft." *Johns Hopkins APL Technical Digest* 32, no. 3 (2013): 558–561.

CIA. 1962. "SD-5 Drone Characteristics and Capabilities." February 14. www.cia.gov/library/readingroom/docs/CIA-RDP71B00822R000100200008-8.pdf.

Clary, David. *The Bekaa Valley—A Case Study*. Maxwell: Air University, 1988.

Corum, James. *The Roots of Blitzkrieg: Hans Von Seeckt and German Military Reform*. Lawrence: Kansas UP, 1992.

Coté, Owen. "The Politics of Innovative Military Doctrine: The U.S. Navy and Fleet Ballistic Missiles." PhD diss., Massachusetts Institute of Technology.

Coulam, Robert. *Illusions of Choice: The F-111 and the Problem of Weapons Acquisition Reform*. Princeton: Princeton UP, 1977.

Doubler, Michael. *Closing with the Enemy: How GIs Fought the War in Europe, 1944–1945*. Lawrence: Kansas UP, 1994.

Ehrhard, Thomas. 2000. "Unmanned Aerial Vehicles in the United States Armed Services: A Comparative Study of Weapon System Innovation." PhD Thesis, Johns Hopkins University.

Ehrhard, Thomas. *Air Force UAV's: The Secret History*. Arlington: Mitchell Institute for Airpower Studies, 2010.

Eliezer, Cohen. *Israel's Best Defense: The First Full Story of the Israeli Air Force*. New York: Orion, 1993.

Emerson, Stephen. *Air War Over North Vietnam: Operation Rolling Thunder-1965–1968*. Barnsley: Pen and Sword, 2018.

Evangelista, Matthew. *Innovation and the Arms Race: How the United States and the Soviet Union Develop New Military Technologies*. Ithaca: Cornell UP, 1988.

Evans, David, and Mark Peattie. *Kaigun: Strategy, Tactics, and Technology in the Imperial Japanese Navy, 1887–1941*. Annapolis: Naval Institute, 1997.

Evron, Yair. *War and Intervention in Lebanon: The Israeli-Syrian Dialogue*. Abingdon: Routledge, 1987.

Fuller, Christopher. *See It, Shoot It: The Secret History of the CIA's Lethal Drone Program*. New Haven: Yale UP, 2017.

GAO. *Improved Strategic Planning Can Enhance DOD's Unmanned Aerial Vehicle Efforts*. Washington DC: USGPO, 2004.

GAO [Government Accounting Office]. *Aquila Remotely Piloted Vehicle Recent Developments and Alternatives*. Washington DC: USGPO, 1986.

Gates, Robert. 2008. *Secretary Gates Remarks at Maxwell-Gunter Air Force Base, Montgomery Alabama*. Transcript: US Department of Defense, April 21. https://archive.defense.gov/Speeches/Speech.aspx?SpeechID=1231 (accessed March, 2020).

Gertler, Jeramiah. *U.S. Unmanned Aerial Systems*. Washington DC: Congressional Research Service, 2012.

Gundmundsson, Bruce. *Stormtroop Tactics: Innovation in the German Army, 1914–1918*. Westport, CT: Praeger, 1995.

Halloran, Richard. 1984. "US Buys Israeli Pilotless Planes." *The New York Times*, May 24.

Harkins, Hugh. *X-45: Uninhabited Combat Air Vehicle*. Glasgow: Centurion, 2013.

Humphreys, Leonard. *The Way of the Heavenly Sword: The Japanese Army in the 1920's*. Palo Alto: Stanford UP, 1995.

Johnson, David. *Fast Tanks and Heavy Bombers: Innovation in the U.S. Army, 1917–1945*. Ithaca: Cornell UP, 1998.

Jones, Christopher. *Unmanned Aerial Vehicles: An Assessment of Historical Operations and Future Possibilities*. Montgomery: Air Command and Staff College Research Paper, 1997.

"Just How Many Predator Drones Does the CIA Have?" 2014. *War Is Boring*, October 15.

Katzenbach Jr, Edward L. "The Horse Cavalry in the Twentieth Century." *Public Policy* 7 (1958): 120–149.

Kier, Elizabeth. *Imagining War: French and British Military Doctrine between the Wars*. Princeton: Princeton UP, 1997.

Kissinger, Henry. 1969. "Air Reconnaissance of South China." October 9. www.cia.gov/library/readingroom/docs/CIA-RDP74J00828R000100200028-9.pdf.

Kuehn, John. *Agents of Innovation: The General Board and the Design of the Fleet that Defeated the Japanese Navy*. Annapolis: Naval Institute, 2008.

LaGrone, Sam. 2019. "U.S. Carriers Need New Lethal Unmanned Aircraft, New Fighter to Stay Relevant." *USNI News*, March 5.

Lambeth, Benjamin. *NATO's Air War for Kosovo: A Strategic and Operational Assessment*. Santa Monica: RAND, 2001.

Mayer, Jane. 2009. "The Predator War: What are the Risks of the C.I.A.'s Covert Drone Program?" *The New Yorker*, October 26.

Merlin, Peter. *Ikhana: Unmanned Aircraft System Western States Fire Missions*. Washington DC: NASA, 2009.

Miles, Donna. 2008. "Gates Forms Task Force to Promote Intelligence, Surveillance for Warfighters." *American Forces Press Service*, April 21.

Miller, Steve. *The 99th Strategic Reconnaissance Squadron: The Air Force's Story of Unmanned Reconnaissance in the Vietnam War*. Everett: Miller Management, 2017.

Newcome, Laurence. *Unmanned Aviation: A Brief History of Unmanned Aerial Vehicles*. Reston: AIAA, 2004.

NRO. 1965. *Annual Report on Activities of the National Reconnaissance Program*, October 31. https://www.cia.gov/library/readingroom/docs/CIA-RDP72R00410R000200010001-5.pdf.

Peebles, Curtis. *Shadow Flights: America's Secret Air War against the Soviet Union*. Novato: Presidio, 2000.

Posen, Barry. *The Sources of Military Doctrine: France, Britain and Germany between the World Wars*. Ithaca: Cornell UP, 1984.

Rich, Ben, and Leo Janos. *Skunk Works*. Boston: Little, Brown and Company, 1994.

Richelson, Jeffrey. *The Wizards of Langley: Inside the CIA's Directorate of Science and Technology*. Cambridge, MA: Westview, 2001.

Rosen, Stephen. *Winning the Next War: Innovation and the Modern Military*. Ithaca: Cornell UP, 1991.

Rynning, Sten. *Changing Military Doctrine: Presidents and Military Power in Fifth Republic France, 1958–2000*. Boulder: Westview, 2001.

Sevastopulo, Demetri. 2007. "US Military in Dogfight over Drones." *The Financial Times*, August 19.

Shanker, Thom. 2008. "At Odds with Air Force, Army Adds Its Own Aviation Unit." *The New York Times*, June 22.

Sherman, Jason. 2006. "Pentagon Sets Plan for New Bomber." *Inside the Air Force*, January 13.

Stevenson, Beth. 2019. "'Loyal Wingman' Part of the Future of Air Combat." *Defense News*, June 13.

Tanguy, Jean-Marc. *Missions Extrêmes: Le GIGN et l'EPIGN en Operation, 1976–2017*. Paris: H&C, 2017.

Wagner, William. *Lightning Bugs and Other Reconnaissance Drones*. Fallbrook, CA: Armed Forces Journal Books, 1982.

Wagner, William, and William Sloan. *Fireflies and Other UAVs*. Arlington, TX: Aerofax, 1992.

Weiner, Sanford. "Evolution in the Post-Cold War Air Force: Technology, Doctrine and Bureaucratic Politics." In *US Military Innovation since the Cold War: Creation without Destruction*, edited by Harvey Sapolsky, Benjamin Friedman, and Brendan Green, 100–118. Abingdon: Routledge, 2009.

Whittle, Richard. *Predator: The Secret Origins of the Drone Revolution*. New York: Henry Holt, 2014.

Xiaoping, Lu. *The PLA Air Force*. New York: CN Times, 2014.

Yenne, Bill. *Attack of the Drones: A History of Unmanned Aerial Combat*. Saint Paul: Zenith, 2004.

Zisk, Kimberley. *Engaging the Enemy: Organization Theory and Soviet Military Innovation, 1955–1991*. Princeton: Princeton UP, 1993.

Armed, unmanned, and in high demand: the drivers behind combat drones proliferation in the Middle East

Francesco F. Milan and Aniseh Bassiri Tabrizi

ABSTRACT

Current dynamics in UCAV proliferation in the Middle East signal that combat drones have become key strategic enablers for state actors in the region, and are no longer seen as an optional asset. With the development of a multitude of indigenous UCAV projects, and the arrival of Chinese-made armed drones on the international market, military procurement in the Middle East has entered a new phase, in which possessing armed drone capabilities is becoming the norm. This article examines the operational and strategic considerations driving Middle Eastern states' UCAV procurement policies, analysing those countries who have been focusing on armed drones for combat purposes and additional intelligence, surveillance, target acquisition, and reconnaissance (ISTAR) needs. The main drivers behind this trend are the operational and strategic advantages brought about by UCAVs, where the benefits related to cost, reliability, and operational risk mitigation are matched by the increased ability to project power that the platform allows, either through deniability or by making UCAVs available to proxies and allies.

Introduction

Once a niche capability owned exclusively by the world's most advanced militaries, over the past decade unmanned aerial vehicles (UAVs)[1] have witnessed an unmatched growth – both in terms of technological sophistication and diffusion. The Middle East, in particular, has emerged as a leading region in the acquisition, deployment, and, at times, indigenous development of their combat variant, known as unmanned combat aerial vehicles (UCAVs).[2] This article examines the operational and strategic considerations driving Middle Eastern state actors' UCAV procurement policies, analysing emerging regional powers who have been focusing on armed drones for combat purposes and additional intelligence, surveillance, target acquisition, and reconnaissance (ISTAR) needs.

The article's main proposition is that current dynamics in UCAV proliferation in the Middle East signal that combat drones have become key strategic enablers, and are no longer seen as an optional asset. In other words, procurement in the Middle East has entered a new phase, in which possessing armed drone capabilities is becoming the norm: in the eyes of Middle Eastern powers, the acquisition of UCAVs is no longer a matter of 'if', but merely a matter of 'when'.[3]

Benefits and advantages such as substantial cost reductions and game-changing levels of risk reduction and operational tempo have convinced militaries worldwide of the value of UCAVs.[4] Still, in the case of the Middle East, the evolution of current security threats, coupled with the end of the United States' monopoly over advanced drone capabilities, particularly in relation to non-stealth UCAVs,[5] have contributed in dramatically accelerating this trend. Virtually all Middle Eastern countries are dealing with either internal or external security threats, and some of them are dealing simultaneously with both. In most cases, the threat is emerging from non-state actors and, as such, presents a significant degree of asymmetry, making UAVs a highly sought-after asset, due to their operational flexibility, technological refinement, and strategic contribution.

This article will discuss how, faced with a strict control regime over armed drones, Middle Eastern countries have been undeterred in pursuing the procurement of UCAVs, further consolidating the idea that possessing armed drones is a 'must'. Such idea has first manifested itself through the mushrooming of indigenous UCAV programs across the region and has been further solidified by the booming of Chinese armed drones purchases of the past few years.

It will then examine how the operational and strategic advantages UCAVs bring about are an ideal fit for the nature of the security challenges in the region's flashpoints. Cost, reliability, and risk mitigation have played a major role in fuelling the operational need for UCAVs; additionally, influence and deniability emerged as the two fundamental strategic drivers contributing to the countries' ability to achieve power projection.

The procurement dilemma

Until recently, the US and Israel held a strict *de facto* duopoly on the production and export of UCAVs, which prevented Middle Eastern countries from obtaining direct access to such capabilities. The US, who has traditionally relied on a policy of stringent export control on the transfer and subsequent use of armed drones, abides by the 'strong presumption of denial' rule set by the Missile Technology Control Regime (MTCR), which regulates the export of ballistic missiles and drones. Based on this principle, the US refuses requests of UCAVs procurement unless it has a substantive degree of certainty the platforms will not be used against its interests, including reputational damage caused by potential human rights violations, or by the equipment or its technology falling into the hands of terrorist groups.[6]

In principle, this rationale has been motivated by the overarching goal of limiting the circulation of sensitive technologies[7]; in practice, this means that the US has rarely allowed for the export of armed drones: only the United Kingdom, France, and Italy currently have US UCAVs in their inventories.[8] In the Middle East, the implementation of these restrictions has meant that even key US allies such as Turkey (a NATO member), the Kingdom of Saudi Arabia (KSA), and the United Arab Emirates (UAE) have all been left empty-handed, despite their requests for UCAVs going back to almost two decades ago.[9] As for Israel, hostility and mutual distrust with countries in the neighbourhood have precluded its companies from selling armed drones in the region, and have also significantly limited the amount of deals on non-armed drones that have been signed.[10]

Over the years, as Middle Eastern governments had to face both Israel's and the United States' refusal to provide them with armed drones, they were left with only two viable alternatives: embarking upon ambitious domestic projects for the production and procurement of indigenous UCAV, or, more recently, the purchase of armed platforms from China. Some earlier UAV programs were upgraded to add lethal payloads to existing ISTAR platforms, effectively turning them into armed drones.

Indigenous programs have faced the technological challenges posed by the development (or, in the best case, acquisition) of critical components such as remote flight controls, sensors, software, and payloads. This has led the sophistication of domestically produced UCAVs to vary massively, depending on factors such as access to technology and engineering skills, as well as prospective end use. Models on the advanced end of the spectrum have been earmarked for national armed forces and the international market, while less sophisticated (or downright outdated) platforms and technology have been made directly available to proxy forces and allies lacking the manpower and technical expertise to operate advanced platforms autonomously.

With the exception of Israel, in the short term, strategic patience seems to have favoured those countries that only looked into acquiring UCAVs once Chinese platforms were made available: while far from performing anywhere close to high-end armed drones such as the Israeli *Eitan* or the US *MQ-9 Reaper*, Chinese drones exported to the Middle East such as the *Wing Loong I/II*, and the *CH-4*, perform comparatively close to the *Hermes 450*, the *MQ-1 Reaper*, and the *Hermes 900* respectively.[11] On the other hand, domestically developed UCAVs such as the Turkish *Bayraktar TB2* or the Emirates' *Yabhon United 40* are either less performing or less sophisticated.

Developing indigenous UCAV programmes

Of the seven Middle Eastern countries that currently operate armed drones, four have been developing their own domestic UCAV program: Israel, Turkey, Iran, and the UAE.

Israel has been a trailblazer in the field of drones, spearheading the development of modern UCAVs since the 1970's: originally concerned by the need to safeguard its aircrews after the deployment of Soviet antiaircraft capabilities around the Suez Canal, it has consistently invested significant manpower and resources, in tandem with the United States, to develop unmanned platforms.[12]

While until mid-2001 Israeli military simulations were concerned with whether the country's UCAV fleet could single-handedly deal with Syrian tanks, changes in threat priorities (such as the rise of Hamas and Ariel Sharon's decision to expand Israel's targeted killings) inaugurated a new era, with armed drones technologies reconfigured for asymmetric conflict, and Israeli UCAV strikes becoming a daily occurrence.[13] Despite the *de facto* absence of Israel from the UCAV market in the Middle East, until 2014 its drones retained their supremacy in global exports, before Chinese UCAVs started flooding the region.[14] Israel currently operates three main indigenous UCAVs: the *Eitan*, the *Hermes 450*, and the *Hermes 900*, originally developed as an ISTAR platform and subsequently upgraded and equipped with combat capabilities. All three platforms have been employed in operations in the Gaza Strip, in the 2006 Israel-Hezbollah War, in the Syrian civil war, as well as in a series of airstrikes in Sudan carried out between 2009 and 2011.[15]

Turkey has also successfully developed an ambitious program for the deployment of indigenous UCAV capabilities. After relying exclusively on US-made and Israeli (unarmed) platforms for almost two decades, the collapse in diplomatic relations with Israel in 2010,[16] coupled with US Congress reluctance to provide armed drones,[17] have led Turkey to focus its efforts on the development of a domestic drone program. As a result, Turkey now operates its own indigenous UAV and UCAV fleet: the *Bayraktar TB2*'s armed version, considered one of the crown jewels of Turkey's drone fleet, has been operative since 2015; it has been deployed to carry out airstrikes on Turkish soil and Northern Iraq against the Kurdistan Workers' Party (PKK), in Northern Syria as part of Turkish military operations in the region, as well as in Libya, where Turkey has been shipping UCAVs and other military equipment to support Al Sarraj's Government of National Accord.[18]

In a similar manner, the UAE launched its own UCAV program after US Congress rejected its original request for the purchase of *Predator* armed drones in 2002. In fact, even in 2004, when the UAE attempted to use its special forces deployment within NATO's International Security Assistance Force (ISAF) in Afghanistan as hard evidence for the need for armed drones, the US did not budge.[19] By then, however, the UAE was already committed to its indigenous program, which eventually led to the introduction of the *Yabhon United 40* in 2013: its armed version is currently in UAE's Air Force inventory and, since 2018, has also been exported to Algeria.[20]

While the US did not hinder the development of indigenous armed drones programs by Israel, Turkey, and the UAE to various degrees, Iran has managed to establish a domestic armed drones program, integrating its own engineering work with reverse engineering carried out on captured US and Israeli drones, despite decades under US sanctions and an arms embargo. It now relies on at least six different indigenous UCAVs[21] that have been operative at least since 2016, when a drone strike against Syrian anti-regime militias near Aleppo was first reported.[22] Since then, its two main UCAVs, the *Shahed-129* and the *Mohajer-6*, have carried out strikes on Iranian soil, as well as in Iraq and Syria.[23]

Turning to Chinese armed drones

China has further contributed to the breaking of the US–Israel duopoly in the field of UCAVs in the Middle East, and since 2013 has been 'selling the hell out of'[24] its armed drones to countries in the region.[25] China imposes far fewer restrictions on who can buy its armed drones or how these can ultimately be used; in fact, by virtue of not being a member of MTCR, China is not bound by its international standards, and, unlike the United States, does not impose conditional end-user agreements preventing weapon systems' retransfer. Neither did the Asian power's emergence on the UCAV market come as a complete surprise to the US, which as early as 2012, described China's growing spending and research on drones as a 'worrisome trend' and its military significance as 'alarming'.[26]

Iraq spearheaded the arrival of Chinese armed drones in the Middle East with a contract signed in 2015, when it acquired *CH-4* UCAVs from Beijing, after being refused access to US *Reapers*, effectively paving the way for all other deals that followed.[27] Since entering the Iraqi inventory, the *CH-4* has carried out more than 200 air strikes against Islamic State targets in the northwestern regions of the country.[28]

Soon after, other countries followed suit: between 2015 and 2017, the UAE secured two separate agreements with China for the provision of *Wing Loong I* and *II* to supplement its *Yabhon United 40* fleet, which does not perform as well as its Chinese equivalents[29]; as a consequence, the Emirati armed forces rely more extensively on their imported drones, which over the past two years have been deployed consistently both in Yemen and in support of general Haftar's military advance in Libya.[30]

The KSA is also on the list of countries that were refused armed US drones, despite its interest in acquiring *Predator* UCAVs to be deployed in the conflict against Houthi rebels in Yemen.[31] Lacking, at least for the time being,[32] an indigenous armed drones program, the KSA formally entered the UCAV owners' club between 2015 and 2016, after signing an agreement with China for the provision of an unspecified number of armed *Wing Loong II* and *CH*-4 for its Air Force.[33] In addition, in 2017 KSA has also signed an agreement for the establishment of a *CH-4* factory on Saudi soil.[34] Similarly to the UAE, since the reception of the drones, the KSA has been deploying its Chinese-made UCAVs in the Yemen conflict.[35]

By virtue of another deal with China, since 2015 Egypt has been employing its *Wing Loong II* in support of special forces operations against *Wilayat Sinai* (the Egyptian branch of ISIS), both in a combat and ISTAR role.[36] Since 2017, the *Wing Loong II* has also appeared at the Uthman Airbase, some 40 km from the border with Libya, where it is employed in the fight against militants and smuggling activities.[37]

Today, all key state actors in the Middle East possess UCAVs. The urgency dictated by ongoing conflicts and the relevant operational needs related to the asymmetric nature of the threats, coupled with the strategic considerations enabled by the extensive degrees of deniability and power projection that UCAVs allow help understanding why the procurement or development of armed drones have become a priority in the region.

Operational and strategic drivers

Operational drivers

Three main operational factors have influenced UCAV procurement in the region: cost, reliability, and risk mitigation.

Whether or not armed drones are part of the inventory, the economic cost of embarking upon a military campaign against non-state actors is generally considerable, especially when accounting for how such campaigns tend to be protracted in time and, ultimately, linger on.[38]

With UCAVs replacing jet fighters in a strike role, the procurement cost per platform can be reduced up to 10 times,[39] while expenses associated with training, maintenance, and operational costs also see a significant reduction: training via simulators is much more effective (and efficient) in the case of UAVs/UCAVs, and the cost per hour of flight is also a tiny fraction of the cost of operating a jet fighter.[40] In addition, UCAVs are generally equipped to be multirole, with payloads allowing them to carry out combat and/or ISTAR tasks.

Current and recent military engagements of Middle Eastern militaries highlight how the cost-effectiveness of armed drones cannot be understated, particularly in the context of asymmetric threats. For instance, the emergence of the First and Second Intifada, together with the campaigns carried out in Lebanon and the recurrent clashes with Hezbollah and Hamas, mean that the

Israeli armed forces have been dealing with asymmetric threats for at least forty years, with huge costs for Israeli's defence and the broader national economy.[41] In a similar vein, Turkey's fight against PKK's insurgency has been dragging on since the mid-1980's; available estimates put the cost of the first two decades of conflict alone (1984–2005) to 88 USDbn.[42]

Saudi Arabia's current campaign against Houthi rebels in Yemen is estimated to have cost 100 USDbn over three years, and, far from achieving any major military breakthrough, has recently witnessed the withdrawal of a substantive part of UAE's forces, Saudi Arabia's main ally in the campaign.[43] In Egypt, the collapse of Hosni Mubarak's regime in 2011 paved the way for the emergence of Wilayat Sinai (the Islamic State's branch on Egyptian soil),[44] whose insurgency in turn led to the allocation of a 11 USDbn military assistance fund from the US in the period 2011–2019,[45] and has witnessed Israeli military operations in the region to support Egypt's efforts.[46]

With such military campaigns being expensive and uncertain in nature, the integration of UCAVs into Middle Eastern countries' military capabilities has at a minimum mitigated the cost of the campaigns' air component: the loss of a drone, after all, is 'affordable even when military budgets are tight or in small countries'.[47] For instance, the UAE *F-16* jet fighter that crashed in Yemen in 2017 had a price tag of 200 USD m, while the Emirati *Wing Loong* shot down over Yemen in April 2019 will cost 2 USD m to replace[48]; the Saudi *AH-64* attack helicopter lost in action in Yemen in 2018 cost 35 USD m, while China offered to replace the two Saudi Air Force-operated *CH-4* UCAVs that crashed in mid-2017 for free.[49] Turkey has lost a 60 USD m-worth *T-129* attack helicopter over Syria in 2018,[50] while procurement for one *Bayraktar TB2* is estimated to cost less than one-tenth of that.[51]

While these examples show how significant the economic benefit of deploying UCAVs over manned platforms is, particularly as it might contribute in prolonging the force commitment, the cost factor alone does not explain the rush in armed drones procurement. Two additional operational advantages need to be taken into account: reliability (or the combination of persistence, speed, and precision), and risk mitigation.

The asymmetric nature of the threats faced by the Middle Eastern countries under examination also contributes to the urgency of obtaining armed drones. Absent conventional frontlines and military control of the territory, determining opposing forces' movements, locations, and operational networks is a key intelligence challenge, and the task's complexities are further exacerbated by the fact that militants might hide in extremely remote or highly populated areas, both of which present unique sets of challenges for ISTAR and strike tasks.[52]

Persistence, or 'the ability to remain on-task for very extended periods of time',[53] means that UAVs can loiter over a location for much longer than other platforms, reducing the risk of losing contact with the target. More importantly, only UCAVs integrate both loitering ISTAR and strike capabilities into a single platform, minimising the time gap between target acquisition and the potential decision to strike high-profile or time-sensitive targets.[54]

Israel built upon its experience in UCAV strikes during the Second Intifada and made extensive use of armed drones during the 2006 conflict with Hezbollah, and in all targeted operations that followed.[55] In addition, the Egyptian government, struggling to maintain security in the Sinai, has been giving the green light to armed Israeli drones to carry out operations against the jihadist militias of Wilayat Sinai,[56] while Egypt's own *Wing Loong* UCAVs have also been employed in support of Egyptian special forces units deployed in the region.[57]

Similarly, for the past two decades, many Turkish military operations against the PKK have been carried out through a combination of UAV-fed ISTAR to commanders, who then, when possible, would scramble jet fighters or attack helicopters to get on location and strike – a chain of actions that can rarely be contained in the span of a few minutes. With the advent of the *Bayraktar TB2*, Turkey has become increasingly reliant on UCAV strikes – both in urban and rural environments on Turkish and Syrian soil, and on the mountains along the Turkish-Iraqi border. The *Bayraktar TB2*, in fact, has become the Turkish military's 'most substantive strategic asset in its counter-terrorism operations', as, among other things, it has significantly reduced the operational tempo between the target acquisition phase and the strike.[58]

While generally there will be *manned* platforms capable of carrying out operations in a satisfactory manner, they will do so at a higher cost, with a slower operational tempo, and with more risks for the units involved. *Unmanned* operations, on the other hand, mitigate those risks, as they 'pose little to no direct risk of harm to operators'.[59] This has very important operational ramifications for countries possessing UCAVs: as operators are remotely located and physically distant from the frontline (as opposed to pilots who have to physically fly into enemy airspace), the only risk to their own safety from enemy action is in case of an attack against the ground control station (GCS) where they are located.

From an operational perspective, the reduction in risks for aircrews means that decision-makers' appetite for risk increases[60]; operations that would have been shelved due to the potential dangers to the pilots' safety can be carried out by drones, as the worst-case scenario involves losing a (relatively cheap) piece of equipment, rather than dealing with killed, captured, or missing aircrews, and with the loss of platforms with hefty price tags.

Non-state actors, however, generally do not possess air power assets; in these cases, the operational risk of air sorties is quite low. It is when adversaries possess air defence system (ADS) capabilities that the risk-mitigating impact of deploying armed drones is particularly significant, and the conflict in Yemen offers an insightful example of this.

Since 2017, Houthi rebels ramped up their ADS capabilities and anti-aircraft activities, with the number of attacks against the Saudi-UAE coalition jet fighters increasing significantly.[61] Both the KSA and UAE adapted their operations accordingly: while the KSA is now relying more extensively on its *CH-4* and *Wing Loong* squadrons flying out of two air bases located along the Yemeni border,[62] the UAE, in line with the strong casualty-aversion dictated by the small size of its force, has also increased its reliance on UCAV-centric operations,[63] and ultimately opted to focus on operations against Al Qaeda in the Arabian Peninsula and strikes against high-profile targets such as the killing of senior Houthi leader Saleh al-Samad in 2018,[64] rather than the broader military campaign in Yemen.[65]

Strategic drivers

Besides operational aspects, the impact UCAVs have on risk dynamics also has important strategic ramifications. As highlighted in a 2015 UN Report:

> "[Armed drones] can alter incentives for the use of force by lowering the risk to one's own armed forces and civilian population or by enabling new types of low-intensity conflict. Their capabilities enable armed forces to pursue new types of missions, tempting States to interpret international humanitarian and human rights law in ways that permit expanded use of force. They are attractive to covert armed forces that operate in ways that may not permit sufficient transparency and accountability."[66]

This trend in military power projection manifests itself via two elements: *influence*, by providing armed drones to allies in conflicts abroad, and *deniability*, through the deployment of UCAVs in a covert manner.

Influence can be enhanced by making armed drones available as a third-party actor to a conflict between two sides (either by selling them, or as part of direct military assistance), giving drone providers more agency on conflict dynamics without risking direct attribution and involvement, while creating a virtuous circle where influence, combined with the faction on the ground's dependency on external support, fuel each other.

Deniability, on the other hand, exploits ambiguity and delayed adversaries' responses (the '*grey zone*' factor).[67] Short-of-war strategies allow for UCAV operations to contribute to 'salami tactics'[68] in which successful action will not be destructive enough to warrant a military response by the adversary, while maintaining a sound degree of deniability will mean attribution might also be challenging, allowing the acting side to keep international scrutiny at bay. Moreover, in case of failure, 'even if these systems are intercepted by States on opposing sides of the conflict, armed UAV users would not lose their own military personnel, and thus feel pressured to retaliate or otherwise escalate' – an unachievable scenario in case manned platforms were the only option available.[69]

While most of the case studies discussed in the article combine, to different degrees, influence and deniability, Turkey's deployment of armed drones constitutes one of the few cases in which influence relies on attribution, rather than deniability, to maximise power projection.

In June 2019, Turkish President Recep Tayyip Erdogan openly stated that his country provides military support to the Libyan Government of National Accord (GNA) to restore the military balance in the conflict, in a direct response to UAE's deployment of armed drones to the Libyan National Army (LNA).[70] Accordingly, Turkey's *Bayraktar TB2* has been spotted carrying out operations over Libya since mid-2019, allowing GNA forces to launch, for the first time, an air strike against LNA's airbase in Jufra, an asset of strategic importance in the advance towards Tripoli.[71]

More frequently, UCAVs have been deployed to exploit plausible deniability, with varying degrees of success. While the existence of enhanced security cooperation between Israel and Egypt is a public affair, Israeli drones operating in the Sinai are unmarked, at times their flight paths are planned so that their location of origin is unclear, and 'alleged' drone strikes are regularly denied by Israeli authorities[72]; in a similar manner, responsibility for an operation against a munitions factory in Sudan in 2012 was neither denied nor confirmed by Israeli authorities.[73] In these cases, by being able to sidestep direct attribution for the operations carried out, Israel has managed to avoid scrutiny regarding the increasingly active role its military forces play in Sinai, its growing diplomatic vicinity to the current Egyptian leadership, as well as on the fact that the conflict against Hamas has quietly expanded into Sudan.

Similar dynamics have been at play in Libya's civil war. A 2017 UN report confirmed that Chinese armed drones have been carrying out operations in Libya since 2016, 'most probably [operated by] the United Arab Emirates'[74]; further UN investigations in 2019 highlighted 'the probable use of Wing Loong UAV variants by the LNA or by a third party in support of the LNA' in violation of the arms embargo[75]; as UN monitoring activities unveiled a range of manned platforms that found their way to the Libyan battlefront from UAE arsenals, there are strong suspicions that the Wing Loong operating in the conflict share the same origins, but no positive identification has been so far achieved.[76]

While countries such as the UAE and Turkey are allowed a degree of leeway in their strategic use of drones, Iran's difficult international position, coupled with its increasingly active role in the region, make deniability a critical diplomatic lifeline: asymmetrical warfare and the ability to use armed drones allows Iran to operate in skies regularly patrolled by Western and Israeli forces.[77] While Iranian authorities have consistently claimed there are no Iranian units operating in Syria, with the exclusion of some 'volunteers' and advisors,[78] over the years Iranian UCAVs have become a regular feature within Syrian airspace.[79] In addition, in April 2019 an Israeli air strike hit the T-4 airbase, located in Homs, a facility of strategic importance for the Syrian regime;[80] 14 Iranian military officers were killed, including the Islamic Revolutionary Guard Corps (IRGC) commander of an Iranian UCAV unit and one of his subordinates.[81] Iran has also been providing armed drones to its allies in the region: while Hezbollah and pro-Iranian Iraqi militias have been receiving armed drones and operational training at least since 2014,[82] more recently Yemen's Houthis have been operating low-tech Iranian drones, which have been subsequently modified to carry out strikes such as the one that killed Yemen's head of military intelligence in January 2019,[83] and the attacks against oil pumping stations located on Saudi territory that were launched in May 2019.[84]

Still, it was the drone air strike against two Aramco oil facilities in Saudi Arabia carried out in mid-September 2019 that showcased UCAV attacks' full potential for deniability and ambiguity.[85] After the Houthis claimed full responsibility for the attack, initial Saudi and American investigations into the events failed to pinpoint the culprit, or even the location from which the attack was launched: attempts to demonstrate Iran's direct responsibility in the attack failed to provide satisfactory and definitive evidence, and even a Saudi Ministry of Defence press conference where the wreckage of the drones used in the attack was put on display only managed to prove that the platforms were Iranian models – but not necessarily Iran-*operated* UCAVs.[86] The incident also highlighted an emerging issue in armed drones proliferation: in a market where buyers are numerous and suppliers very limited, it becomes harder to identify users involved in an attack. Models and their provenance will provide little granularity regarding attribution: additional technical analysis on elements such as modifications, flight paths, and components might bring additional clarity, granted that these are available to be examined.

Even from a strategic perspective, armed drones have significantly expanded Middle Eastern countries' ability to project their power in a cost-effective manner. The degree of deniability the use of UCAVs offers has been exploited significantly, as the frequency of unattributed or undetected drone operations shows; in a similar manner, in the changing geopolitical landscape of the region, the opportunity to expand one's influence with limited direct risks has made armed drones the go-to capability to support like-minded factions in the region.

A game-changing capability

The development of indigenous UCAV projects, coupled with China's entry into the armed drones export business, have signalled the end of the US-Israeli duopoly that until recently governed the global armed drones market.

Such paradigmatic change has nullified the *de facto* veto power the two countries had on allowing direct access to armed drones to military forces worldwide. Despite the (now) available alternatives, most Middle Eastern actors would still rather buy American, due to the superior technological quality, reliability, and precision US-made UCAVs offer, as well as the potential for integration with other systems. Over time, however, strategic priorities have taken precedence, and all key Middle Eastern military powers, faced with no alternative options, have opted to 'go solo' or buy Chinese UCAVs.[87]

As an additional asset to Middle Eastern military forces, armed drones have been procured due to their unmatched reliability and precision, as well as for their direct effect on improving the 'blood and treasure' cost of military campaigns.[88]

But the need to obtain armed drones has also been dictated by the manifest strategic advantages that such platforms bring to the table. These drivers have, in many cases, fed into an underlying change in strategic priorities that many Middle Eastern governments have been pursuing. If, on the one hand, armed drones are a cost-effective and operationally savvy alternative to manned options, on the other hand, they also bear the potential for an increase in the appetite for strategic risk.

Armed drones, in other words, broaden the spectrum of strategic options available to the countries owning them, truly becoming a strategic enabler and an instrument of power projection for those governments who might be seeking to expand their regional role. Testament to this principle is not just UCAVs' increased presence across battlefields in the Middle East, but, more importantly, the fact that armed drones have rapidly become the go-to asset to be provided to proxy actors, allies, and like-minded forces: the UAE–Turkey drone-centric tit-for-tat approach to the conflict in Libya echoes the provision of Iranian unmanned assets to Houthi militias in Yemen and to allied forces in Syria and Iraq.

While the type of support provided to partner factions in conflict includes a vast array of capabilities, because of their operational and strategic edge UCAVs can act as a significant force multiplier for those actors who want to project their power in the region while maintaining a degree of separation from the actions on the ground: operations such as the strikes deep into Saudi territory claimed by Houthi authorities would have been considered unachievable without an external boost in the form of unmanned combat assets; more importantly, attribution has been highly challenging, guaranteeing a degree of impunity for the perpetrators.

As much as Israel and the United States, pioneers in the development and employment of armed drones, have progressively become more proactive, dependent on and inventive in the use of armed drones, a new generation of UCAV owners is now dealing with a similar operational learning curve. Not unlike their predecessors, they are also experiencing the strategic opportunities that possessing such capability endows.

Notes

1. UAVs are also commonly referred to as 'drones'. This also includes the sub-category of *armed* drones, known as unmanned combat aerial vehicles (UCAVs).
2. New America Foundation, "World of Drones."
3. Fuhrmann and Horowitz, "Droning On."
4. Wills, *Unmanned Combat Air Systems in Future Warfare*, 17.
5. Horowitz et al., "Separating Fact from Fiction in the Debate over Drone Proliferation," 41; Sayler, *A world of proliferated drones*, 24–26.
6. Missile Technology Control Regime (MTCR) Annex Handbook, ii–iii; Zenko and Kreps, *Limiting Armed Drones Proliferation*, 17–19.
7. Weitz, "Reforming US export control reforms," 19.
8. Weingberger, "China Has Already Won the Drone Wars," *Foreign Policy*. In April 2018, the Trump administration revised the US policy on the export of armed drones, but so far this reform has not removed the 'strong presumption of denial' rule for selling the more advanced systems, thus leading to a lack of concrete results. Trump, *National Security Presidential Memorandum Regarding U.S. Conventional Arms Transfer Policy*; Seligman, 'Trump's Push to Boost Lethal Drone Exports Reaps Few Rewards', *Foreign Policy*.
9. Detsch, "Pentagon warns Middle East turning to China for drone needs", *Al Monitor*; Manson, "Trump relaxes conditions over sale of US killer drones", *Financial Times*.
10. Boyle, "The Race for Drones November," 81.
11. "Is China at the Forefront of Drone Technology?," *CSIS ChinaPower*.
12. Clark, "Uninhabited Combat Aerial Vehicles," 22–27.
13. Bergman, *Rise and Kill First*, xxi; 455–60.
14. Boyle, "The Race for Drones," 81; Wezeman et al. *Trends in International Arms Transfers*. SIPRI.
15. Bassiri Tabrizi and Bronk, *Armed Drones in the Middle East: Proliferation and Norms in the Region*, 16; Cordesman et al., *Lessons of the 2006 Israel-Hezbollah War*, 106–107; Issacharov, "Report: Hamas Arms Trafficker Killed in Sudan Air Strike," *Haaretz;* How Israel Foiled an Arms Convoy Bound for Hamas, *Time*.
16. "Israeli Manufacturers' Turkish UAV Contract."
17. Farooq, "The Second Drone Age."
18. Gauthier-Villars and Maslin, 'In Libyan War, Turkey Takes Sides Against Mideast Rivals."
19. Frew, *Drone Wars: The Next Generation*, 20; Wikileaks Cable #04ABUDHABI2113, "Shaykh Mohamed bin Zayed Rejects Unarmed Predator Proposal," U.S. Embassy Abu Dhabi.
20. Jennings, "Algeria shown to be operating UAE-developed UAVs."; Boyle, "The Race for Drones," 84.
21. O'Gorman and Abbott, "Remote control war – Unmanned combat air vehicles in China, India, Israel, Iran, Russia and Turkey," 9.
22. Mcleary, "Iranian Drones Now Hitting Rebel Targets in Syria."
23. Borger, "US shoots down second Iran-made armed drone over Syria in 12 days."
24. Turak, "Pentagon is scrambling as China "sells the hell out of" armed drones to US allies,"
25. Bassiri Tabrizi and Bronk, *Armed Drones in the Middle East*; "The growing appetite for armed drones in the Middle East."
26. US Department of Defense, *The Role of Autonomy in DoD Systems*, 70.

27. Marcus, "China helps Iraq military enter drone era."; Gordon and Schmitt, "U.S. Said to Rebuff Iraqi Request to Strike Militants."
28. Delalande, "China's Drones Are Taking on ISIS (And Here's Why America Should Worry)."
29. Biggers, "UAE United 40 Block 5 at Test Airfield."
30. United Nations Security Council, S/2017/466, 144.
31. Wikileaks Cable #10RIYADH159_a, "Saudi Arabia: Renewed assurances on satellite imagery."
32. In February 2019, Saudi company Science Technology announced the launch of development program for a new UCAV platform. See "Science Technology launches program to develop heavy unmanned bomber."
33. Minnick, "China Again Tries To Pierce Gulf Defense Market."
34. Chan, "Chinese drone factory in Saudi Arabia first in Middle East."
35. United Nations Security Council, S/2017/466, 144.
36. Page and Sonne, "Unable to buy U.S. military drones, Allies place orders with China."
37. Gettinger, "Drone Bases Updates."
38. Arreguin-Toft, "How the Weak Win Wars," 103–5.
39. Boyle, "The Race for Drones," 79.
40. Wills, *Unmanned Combat Air Systems in Future Warfare*, 29; and Fallows, "The Tragedy of the American Military."
41. Plocker, *The landscape of the Israeli economy, society and policy after 1200 days of Intifada*.
42. Mutlu, "The Economic Cost of Civil Conflict in Turkey."
43. Guzanski and Heinstein, "Saudi Arabia's War in Yemen Has Been a Disaster."; and Walsh and Kirkpatrick, "U.A.E. Pulls Most Forces From Yemen in Blow to Saudi War Effort."
44. Heller, Christian H. "The Sinai Insurgency."
45. Congressional Research Service, "Egypt: Background and U.S. Relations," 27.
46. Kirkpatrick, "Secret Alliance."
47. Boehler and Doyle, "Use by Iraqi Military May Be a Boon for China-Made Drones."
48. Aboufade, "UAE fighter jet crashes in Yemen, 2 pilots killed."; and Demerly, "Let's Talk About The Photo of Chinese-Built "Wing Loong" Drone (Likely Operated by UAE) over Libya."
49. Interview with Saudi military officer, June 2019; "Arab coalition pilot and assistant killed in helicopter crash in Yemen."; US Office of the Under Secretary of Defense/Chief Financial Officer, *Program Acquisition Cost by Weapon System*, March 2014.
50. Interview with Turkish Undersecretariat for Defence Industries senior official, July 2011.
51. Guntay Simsek, "Ilk milli IHA kac paraya mal oldu?"
52. Kittner, "The Role of Safe Havens in Islamist Terrorism."
53. House of Commons Defence Committee, *The contribution of Unmanned Aerial Vehicles to ISTAR capability*, 17.
54. Zenko and Kreps, "Limiting Armed Drones Proliferation," 8–9.
55. Dobbing and Cole, *Israel and the Drone War*, 11–13; and "Precisely Wrong: Gaza Civilians Killed by Israeli Drone-Launched Missiles."
56. Kirkpatrick, "Secret Alliance."
57. An ISIS propaganda video shows an armed *Wing Loong* over Sinai – https://twitter.com/mahmouedgamal44/status/1063142938161672192.

58. Gurcan, "In struggle against PKK, Turkey takes flight."
59. Woodhams and Borrie, "Armed UAVs in conflict escalation and inter-State crisis," 1.
60. Zenko and Kreps, "Limiting Armed Drones Proliferation," 8.
61. Nadimi and Knights, "Iran's Support to Houthi Air Defenses in Yemen."; and Bassiri Tabrizi and Bronk, *Armed Drones in the Middle East*, 21.
62. Interview with Saudi military officer, September 2018; https://twitter.com/Obs_IL/status/1151493541685800960.
63. Interview with Western senior diplomat, Abu Dhabi, May 2018; Interview with UAE security expert, Abu Dhabi, May 2018.
64. Shaif and Watling, "How the UAE's Chinese-Made Drone Is Changing the War in Yemen."
65. Wintour and McKernan, "Yemen: UAE confirms withdrawal from port city of Hodeidah."
66. United Nations, *Study on Armed Unmanned Aerial Vehicles Prepared on the Recommendation of the Advisory Board on Disarmament Matters.*
67. Hicks et al., "Zone Defense," 3–5.
68. Wirtz, "Life in the 'Gray Zone'," 107–9.
69. Woodhams and Borrie, "Armed UAVs in conflict escalation and inter-State crisis", 11; Stohl, "Exercising Restraint? The New U.S. Rules for Drone Transfers."
70. Soylu, "Turkey sold arms and equipment to UN-backed Libyan government, Erdogan confirms."
71. Al-Atrush, "Libya Hit Haftar Forward Airbase After Receiving Turkish Drones."
72. Kirkpatrick, "Secret Alliance"; Hubbard and Kershner, "Sinai Blasts Kill Up to 5 Thought to Be Militants."
73. Black, "Israeli attack."
74. United Nations Security Council, S/2017/466, 144.
75. Nichols, "U.N. report finds likely use of armed drone in Libya by Haftar or 'third party'".
76. United Nations Security Council, S/2017/466, 24–34.
77. Tabatabai, "Containment and Strike: Iran's Drone Program."
78. "Iran's army denies direct involvement in Syria, speaks of 'volunteers'".
79. Nadimi, "Israel Once Again Strikes Iran's UAV Base in Syria."
80. Oryx, "Fortress T4: An Airbase at War."
81. See note 79 above.
82. Dehghanpisheh and Stewart, "From Iraq to Yemen, drones raise U.S. alarm over Iranian plans."; Horowitz, "Separating Fact from Fiction in the Debate over Drone Proliferation," 36.
83. Brumfiel, "In Yemen Conflict, Some See A New Age Of Drone Warfare."
84. "Saudi Arabia oil installations hit in drone attacks."
85. Hubbard et al., "Two Major Saudi Oil Installations Hit by Drone Strike, and U.S. Blames Iran."
86. Schmitt et al., "To Find Clues in Saudi Oil Attacks, U.S. Examines Missile and Drone Parts."; and Rezaian, "The Saudi-Iran rivalry isn't new, but it's getting riskier by the hour."
87. Interviews with UAE security expert and Western senior diplomat Abu Dhabi, May 2018; interview with Saudi military officer, September 2018.
88. Zegart, "Cheap Flights, Credible Threats," 15–21.

Disclosure statement

Both authors declare there has been no conflict of interest in publishing this article.

ORCID

Francesco F. Milan http://orcid.org/0000-0003-3555-5113
Aniseh Bassiri Tabrizi http://orcid.org/0000-0002-0268-0201

Bibliography

Aboufade, Leith. 2017. "UAE Fighter Jet Crashes in Yemen, 2 Pilots Killed". *Almadar News*, October 18. https://www.almasdarnews.com/article/uae-fighter-jet-crashes-yemen-2-pilots-killed/

Al-Atrush, Samer. 2019. "Libya Hit Haftar Forward Airbase After Receiving Turkish Drones". *Bloomberg*, July 26. https://www.bloomberg.com/news/articles/2019-07-26/tripoli-government-says-it-struck-haftar-s-main-forward-airbase.

"Arab Coalition Pilot and Assistant Killed in Helicopter Crash in Yemen". 2018. *Arab News*, September 14. https://www.arabnews.com/node/1372026/saudi-arabia.

Arreguin-Toft, Ivan M. "How the Weak Win Wars: A Theory of Asymmetric Conflict." *International Security* 26, no. 1 (2001): 93–128. doi:10.1162/016228801753212868.

Bassiri Tabrizi, Aniseh, and Justin Bronk. *Armed Drones in the Middle East: Proliferation and Norms in the Region*. London: Royal United Services Institute, 2018.

Bergman, Ronen. *Rise and Kill First: The Secret History of Israel's Targeted Assassinations*. New York, NY: Random House, 2018.

Biggers, Chris. 2016. "UAE United 40 Block 5 at Test Airfield". *Bellingcat*, March 26. https://www.bellingcat.com/news/2015/03/26/uae-united-40-block-5-at-test-air field/.

Black, Ian. 2012. "'Israeli Attack' on Sudanese Arms Factory Offers Glimpse of Secret War". *The Guardian*, October 25. https://www.theguardian.com/world/2012/oct/25/israeli-sudanese-factory-secret-war.

Boehler, Patrick, and Gerry. Doyle 2015. "Use by Iraqi Military May Be a Boon for China-Made Drones". *The New York Times*, December 17. https://www.nytimes.com/2015/12/18/business/international/china-drone-export-iraq.html.

Borger, Julian 2017. "US Shoots down Second Iran-made Armed Drone over Syria in 12 Days". *The Guardian*, June 20. https://www.theguardian.com/us-news/2017/jun/20/us-iran-drone-shot-down-syria.

Boyle, Michael J. "The Race for Drones November." *Orbis* 59, no. 1 (2015): 76–94. doi:10.1016/j.orbis.2014.11.007.

Brumfiel, Geoff. 2019. "In Yemen Conflict, Some See A New Age Of Drone Warfare". *NPR*, May 29. https://www.npr.org/2019/05/29/726760128/in-yemen-conflict-some-see-a-new-age-of-drone-warfare?t=1567700504682.

Chan, Minnie. 2017. "Chinese Drone Factory in Saudi Arabia First in Middle East". *SCMP*, March 26. https://www.scmp.com/news/china/diplomacy-defence/article/2081869/chinese-drone-factory-saudi-arabia-first-middle-east.

Clark, Richard M. *Uninhabited Combat Aerial Vehicles: Airpower by the People, For the People, But Not with the People*. CADRE Paper No.8. Alabama: Air University Press, 2000.

Congressional Research Service. 2019. "Egypt: Background and U.S. Relations". March 12. https://fas.org/sgp/crs/mideast/RL33003.pdf.

Cordesman, Antony H, George Sullivan, and William D. Sullivan. *Lessons of the 2006 Israel-Hezbollah War*. Washington, DC: Center for Strategic and International Studies, 2007.

Dehghanpisheh, Babak, and Phil Stewart. 2019. "From Iraq to Yemen, Drones Raise U.S. Alarm over Iranian Plans". *Reuters*, July 17. https://www.reuters.com/article/us-mideast-iran-drones/from-iraq-to-yemen-drones-raise-u-s-alarm-over-iranian-plans-idUSKCN1UC1X4.

Delalande, Arnaud. 2018. "China's Drones are Taking on ISIS (And Here's Why America Should Worry)". *The National Interest*, February 21. https://nationalinterest.org/blog/the-buzz/chinas-drones-are-taking-isis-heres-why-america-should-worry-24581.

Demerly, Tom. 2019. "Let's Talk about the Photo of Chinese-Built "Wing Loong" Drone (Likely Operated by UAE) over Libya". *The Aviationist*, July 3. https://theaviationist.com/2019/07/03/lets-talk-about-the-photo-of-chinese-built-wing-loong-ii-drone-likely-operated-by-uae-over-libya/.

Detsch, Jack. "Pentagon Warns Middle East Turning to China for Drone Needs." *Al Monitor*, January 16 (2019). https://www.al-monitor.com/pulse/originals/2019/01/pentagon-mideast-turning-china-drone-needs.html.

Dobbing, Mary, and Chris Cole. *Israel and the Drone War*. Oxford: Drone Wars UK, 2014.

Fallows, James. "The Tragedy of the American Military". *The Atlantic*, January 2016. https://www.theatlantic.com/magazine/archive/2015/01/the-tragedy-of-the-american-military/383516/

Farooq, Umar. 2019. "The Second Drone Age: How Turkey Defied the U.S. And Became a Killer Drone Power". *The Intercept*, May 14. https://theintercept.com/2019/05/14/turkey-second-drone-age/.

Frew, Joanna. *Drone Wars: The Next Generation*. Oxford: Drone Wars UK, 2018.

Fuhrmann, Matthew, and Michael C. Horowitz. "Droning On: Explaining the Proliferation of Unmanned Aerial Vehicles." *International Organization* 71, no. 2 (2017): 397–418. doi:10.1017/S0020818317000121.

Gauthier-Villars, David, and Jared Maslin. 2019. "In Libyan War, Turkey Takes Sides Against Mideast Rivals". *The Wall Street Journal*, July 2. https://www.wsj.com/articles/in-libyan-war-turkey-takes-sides-against-mideast-rivals-11562065082.

Gettinger, Dan. "Drone Bases Updates". https://dronecenter.bard.edu/drone-bases-updates/

Gordon, Michael R., and Eric Schmitt. "U.S. Said to Rebuff Iraqi Request to Strike Militants." *The New York Times*. https://www.nytimes.com/2014/06/12/world/middleeast/iraq-asked-us-for-airstrikes-on-militants-officials-say.html.

"The Growing Appetite for Armed Drones in the Middle East". 2019. *The Economist*, March 9. https://www.economist.com/middle-east-and-africa/2019/03/09/the-growing-appetite-for-armed-drones-in-the-middle-east.

Gurcan, Metin. 2016. "In Struggle against PKK, Turkey Takes Flight". *Al Monitor*, October 14. https://www.almonitor.com/pulse/fr/contents/articles/originals/2016/10/turkey-pkk-learn-from-isis-drones.html.

Guzanski, Yoel, and Ari. Heinstein 2018. "Saudi Arabia's War in Yemen Has Been a Disaster". *The National Interest*, March 25. https://nationalinterest.org/feature/saudi-arabias-war-yemen-has-been-disaster-25064.

Heller, "The Sinai Insurgency: The Next ISIS Crisis?" *Small Wars Journal*. https://smallwarsjournal.com/jrnl/art/sinai-insurgency-next-isis-crisis

Hicks, Kathleen H., John Schaus, and Michael. Matlaga. *Zone Defense: Countering Competition in the Space between War and Peace*. Washington, DC: Center for Strategic and International Studies, 2018. November.

Horowitz, Michael, Sarah Kreps, and Matthew Furhmann. "Separating Fact from Fiction in the Debate over Drone Proliferation." *International Security* 41, no. 2 (2016): 7–42. doi:10.1162/ISEC_a_00257.

House of Commons Defence Committee. 2008. *The contribution of Unmanned Aerial Vehicles to ISTAR capability*, July 15. https://publications.parliament.uk/pa/cm200708/cmselect/cmdfence/535/535.pdf

How Israel Foiled an Arms Convoy Bound for Hamas. 2009. *Time*, March 30. http://content.time.com/time/world/article/0,8599,1888352,00.html.

Hubbard, Ben, Palko Karasz, and Stanley Reed. 2019. "Two Major Saudi Oil Installations Hit by Drone Strike, and U.S. Blames Iran". *The New York Times*, September 14. https://www.nytimes.com/2019/09/14/world/middleeast/saudi-arabia-refineries-drone-attack.html?module=inline.

Hubbard, Ben, and Isabel Kershner. 2013. "Sinai Blasts Kill up to 5 Thought to Be Militants". *The New York Times*, August 9. https://www.nytimes.com/2013/08/10/world/middleeast/sinai-blasts-kill-up-to-5-islamic-militants.html?module=inline

"Iran's Army Denies Direct Involvement in Syria, Speaks of 'Volunteers'". 2016. *Reuters*, April 20. https://uk.reuters.com/article/uk-mideast-crisis-syria-iran-idUKKCN0XH11J.

"Is China at the Forefront of Drone Technology?". 2018. *CSIS ChinaPower*, May 29. https://chinapower.csis.org/china-drones-unmanned-technology/.

"Israeli Manufacturers' Turkish UAV Contract". 2011. *Defense Industry Daily*, December 22. https://www.defenseindustrydaily.com/israeli-manufacturers-win-150m-turkish-uav-contract-updated-0389/.

Issacharov, Avi. 2011. "Report: Hamas Arms Trafficker Killed in Sudan Air Strike". *Haaretz*, April 6. https://www.haaretz.com/1.5148045.

Jennings, Gareth. "Algeria Shown to Be Operating UAE-developed UAVs". *Jane's Defence: Air Platforms*. 21 December 2018. https://www.janes.com/article/85378/algeria-shown-to-be-operating-uae-developed-uavs.

Kirkpatrick, David D. 2018. "Secret Alliance: Israel Carries Out Airstrikes in Egypt, with Cairo's O.K." *The New York Times*, February 3. https://www.nytimes.com/2018/02/03/world/middleeast/israel-airstrikes-sinai-egypt.html.

Kittner, Cristiana C. Brafman. "The Role of Safe Havens in Islamist Terrorism." *Terrorism and Political Violence* 19, no. 3 (2007): 307–329. doi:10.1080/09546550701246791.

Manson, 2018. "Trump Relaxes Conditions over Sale of US Killer Drones". *Financial Times*, April 19. https://www.ft.com/content/a48b8f56-43e4-11e8-803a-295c97e6fd0b.

Marcus, Jonathan. 2015. "China Helps Iraq Military Enter Drone Era". *BBC News*, October 12. https://www.bbc.co.uk/news/world-middle-east-34510126.

Mcleary, Paul. 2016 "Iranian Drones Now Hitting Rebel Targets in Syria". *Foreign Policy*, February 29. https://foreignpolicy.com/2016/02/29/iranian-drones-now-hitting-rebel-targets-in-syria/.

Minnick, Wendell. 2015. "China Again Tries To Pierce Gulf Defense Market", *Defense News*, November 6. https://www.defensenews.com/thought-leadership/digital-show-dailies/dubai-air-show/2015/11/06/china-again-tries-to-pierce-gulf-defense-market/.

Missile Technology Control Regime (MTCR) Annex Handbook. 2017. http://mtcr.info/wordpress/wp-content/uploads/2017/10/MTCR-Handbook-2017-INDEXED-FINAL-Digital.pdf

Mutlu, Servet. "The Economic Cost of Civil Conflict in Turkey." *Middle Eastern Studies* 47, no. 1 (2011): 63:80. doi:10.1080/00263200903378675.

Nadimi, Farzin. 2018. "Israel once Again Strikes Iran's UAV Base in Syria". *The Washington Institute for Near East Policy*, April 10. https://www.washingtoninstitute.org/policy-analysis/view/israel-once-again-strikes-irans-uav-base-in-syria.

Nadimi, Farzin, and Michael. Knights 2018. "Iran's Support to Houthi Air Defenses in Yemen". *The Washington Institute for Near East Policy*, April 4. https://www.washingtoninstitute.org/policy-analysis/view/irans-support-to-houthi-air-defenses-in-yemen.

New America Foundation. "World of Drones". 2017. https://www.newamerica.org/in-depth/world-of-drones/.

Nichols, Michelle. 2019. "U.N. Report Finds Likely Use of Armed Drone in Libya by Haftar or 'Third Party'". *Reuters*, May 8. https://af.reuters.com/article/commoditiesNews/idAFL2N22K1G3?fbclid=IwAR00IkfvtaDr-ClqVLAy-cbXLZqP26XRG5HN21PhSDVbfR4rPG1jenWPl7g

O'Gorman, Rob, and Chris Abbott. *Remote Control War Unmanned Combat Air Vehicles in China, India, Israel, Iran, Russia and Turkey*. Bloomsbury: Open Briefing, 2013.

Office of the Under Secretary of Defense/Chief Financial Officer. 2014. *Program Acquisition Cost by Weapon System*. March. https://comptroller.defense.gov/Portals/45/documents/defbudget/fy2015/fy2015_Weapons.pdf

Oryx. 2015. "Fortress T4: An Airbase at War". *Bellingcat*, June 29. https://www.bellingcat.com/news/mena/2015/06/29/fortress-t4-an-airbase-at-war/.

Page, Jeremy, and Paul. Sonne 2017. "Unable to Buy U.S. Military Drones, Allies Place Orders with China." *Wall Street Journal*, July 17. https://www.wsj.com/articles/unable-to-buy-u-s-military-drones-allies-place-orders-with-china-1500301716.

Plocker, Sever. "The Landscape of the Israeli Economy, Society and Policy after 1200 Days of Intifada." https://www.brookings.edu/wp-content/uploads/2012/04/20031113.pdf.

"Precisely Wrong: Gaza Civilians Killed by Israeli Drone-Launched Missiles". 2009. *Human Rights Watch*, June 30. https://www.hrw.org/report/2009/06/30/precisely-wrong/gaza-civilians-killed-israeli-drone-launched-missiles.

Rawan, Shaif, and Jack Watling. 2018. "How the UAE's Chinese-Made Drone Is Changing the War in Yemen". *Foreign Policy*, April 27. https://foreignpolicy.com/2018/04/27/drone-wars-how-the-uaes-chinese-made-drone-is-changing-the-war-in-yemen/

Rezaian, Jason. 2019. "The Saudi-Iran Rivalry Isn't New, but It's Getting Riskier by the Hour". *The Washington Post*, September 18. https://www.washingtonpost.com/opinions/2019/09/18/saudi-iran-rivalry-isnt-new-its-getting-riskier-by-hour/.

Sayler, Kelley. 2015. *A world of proliferated drones: A technology primer.* Center for a New American Security. June. https://www.files.ethz.ch/isn/191911/CNAS%20World%20of%20Drones_052115.pdf

Schmitt, Eric, Julian E. Barnes, and David D. Kirkpatrick 2019. "To Find Clues in Saudi Oil Attacks, U.S. Examines Missile and Drone Parts". *The New York Times*, September 17. https://www.nytimes.com/2019/09/17/world/middleeast/iran-attacks-saudi-oil.html.

"Science Technology Launches Program to Develop Heavy Unmanned Bomber." 2019. *Defense Blog*, March 13. https://defence-blog.com/news/science-technology-launches-program-to-develop-heave-unmanned-bomber.html?fbclid=IwAR1gebqF-QDERJfVVdUg0SA7A-.

Seligman, Lara. "Trump's Push to Boost Lethal Drone Exports Reaps Few Rewards". 2018. *Foreign Policy*, December 6. https://foreignpolicy.com/2018/12/06/trump-push-to-boost-lethal-drone-exports-reaps-few-rewards-uas-mtcr/2v6pmnEiR2OnBBbscT-ENrTmDILvs44wI.

Soylu, Ragip. 2019. "Turkey Sold Arms and Equipment to UN-backed Libyan Government, Erdogan Confirms". *Middle East Eye*, June 20. https://www.middleeasteye.net/news/turkey-sold-equipment-libya-erdogan-confirms.

Stohl, Rachel. 2015. "Exercising Restraint? The New U.S. Rules for Drone Transfers". *Arms Control Association*, May. https://www.armscontrol.org/act/2015-05/features/exercising-restraint-new-us-rules-drone-transfers.

Tabatabai, Ariane. "Containment and Strike: Iran's Drone Program." *Terrorism Monitor* 15, no. 17 (2017). https://jamestown.org/program/containment-and-strike-irans-drone-program/.

Trends in International Arms Transfers. Stockholm: Stockholm International Peace Research Institute, 2019.

Trump, Donald J. 2019. *National Security Presidential Memorandum Regarding U.S. Conventional Arms Transfer Policy*, April 19. https://www.whitehouse.gov/presidential-actions/national-security-presidential-memorandum-regarding-u-s-conventional-arms-transfer-policy/.

Turak, Natasha. "Pentagon Is Scrambling as China 'Sells the Hell Out Of' Armed Drones to US Allies". 2019. *CNBC*, February 21. https://www.cnbc.com/2019/02/21/pentagon-is-scrambling-as-china-sells-the-hell-out-of-armed-drones-to-americas-allies.html.

United Nations. 2015. "Study on Armed Unmanned Aerial Vehicles Prepared on the Recommendation of the Advisory Board on Disarmament Matters." https://unoda-web.s3-accelerate.amazonaws.com/wp-content/uploads/assets/publications/more/drones-study/drones-study.pdf.

United Nations Security Council. S/2017/466. https://www.securitycouncilreport.org/atf/cf/%7B65BFCF9B-6D27-4E9C-8CD3-CF6E4FF96FF9%7D/s_2017_466.pdf

US Department of Defense. *The Role of Autonomy in DoD Systems*. Washington, D.C: Defense Science Board, 2012.

Walsh, Declan, and David D. Kirkpatrick 2019. "U.A.E. Pulls Most Forces from Yemen in Blow to Saudi War Effort". *The New York Times*, July 11. https://www.nytimes.com/2019/07/11/world/middleeast/yemen-emirates-saudi-war.html.

Weingberger, Sharon. 2018. "China Has Already Won the Drone Wars". *Foreign Policy*, May 10. https://foreignpolicy.com/2018/05/10/china-trump-middle-east-drone-wars/.

Weitz, Richard. 2015. *Reforming US Export Control Reforms: Advancing US Army Interests*. Strategic Studies Institute and U.S. Army War College Press, December. https://publications.armywarcollege.edu/pubs/2376.pdf.

Wezeman, Pieter D, Aude Fleurant, Alexandra Kuimova, Nan Tian, and T. Wezeman Siemon

Wikileaks Cable #04ABUDHABI2113. 2004. "Shaykh Mohamed Bin Zayed Rejects Unarmed Predator Proposal". *U.S. Embassy Abu Dhabi*. June 27.

Wikileaks Cable #10RIYADH159_a. 2010. "Saudi Arabia: Renewed Assurances on Satellite Imagery." *U.S. Embassy Riyadh*, February 7.

Wills, Colin. *Unmanned Combat Air Systems in Future Warfare: Gaining Control of the Air*. Hampshire: Palgrave MacMillan, 2015.

Wintour, Patrick, and Bethan McKernan. 2019. "Yemen: UAE Confirms Withdrawal from Port City of Hodeidah". *The Guardian*, July 9. https://www.theguardian.com/world/2019/jul/09/yemen-uae-confirms-withdrawal-from-port-city-of-hodeidah

Wirtz, James J. "Life in the 'Gray Zone': Observations for Contemporary Strategists." *Defense & Security Analysis* 33, no. 2 (2017): 106–114. doi:10.1080/14751798.2017.1310702.

Woodhams, George, and John Borrie. 2018. "Armed UAVs in Conflict Escalation and inter-State Crisis". United Nations Institute for Disarmament Research. http://www.unidir.org/files/publications/pdfs/armed-uav-in-conflict-escalation-and-inter-state-crisis-en-747.pdf.

Zegart, Amy. "Cheap Flights, Credible Threats: The Future of Armed Drones and Coercion." *Journal of Strategic Studies* 43, no. 1 (2020): 6–46. doi:10.1080/01402390.2018.1439747.

Zenko, Micah, and Sarah. Kreps. *Limiting Armed Drones Proliferation*. New York: Council on Foreign Relations Press, 2014.

U.S. drone campaign in Pakistan's Pashtun 'tribal' region: beginning of the end under President Trump?

Farooq Yousaf

ABSTRACT

The U.S. drone campaign in Pakistan's Pashtun 'tribal' areas, formerly known as FATA, has remained one of the most contentious issues since the war on terror began in 2001 in Afghanistan. Though drone strikes have polarised opinions, they have also resulted in the elimination of lower, middle and higher-level leaders of all militant groups based in the 'tribal' region. The U.S. drone campaign in the former-FATA region had some success in the elimination of terrorists. But lack of access to the 'tribal' areas coupled with the secrecy surrounding the drone campaign resulted in major discrepancies in the total number of U.S. drone strikes and the resultant casualties. These factors have contributed towards an anti-drone and anti-U.S. narrative in Pakistan, largely manufactured by Pakistani mainstream media. With the last drone strike taking place almost a year ago, this article argues that the U.S. drone campaign may have finally halted in Pakistan. However, even with a gradual halt, the 'tribal' areas can witness drone strikes in the future if high-profile terrorists, based on actionable intelligence, are located in the region.

Introduction

When the U.S.-led coalition invaded Afghanistan in October 2001, the resulting war on terror also brought direct and indirect consequences to Pakistan. One direct consequence was Pakistan's Pashtun 'tribal' areas (formerly known as the Federally Administered Tribal Areas, or FATA, see below Figure 1) becoming a direct target of the U.S. drone campaign that formally started in 2004. The erstwhile-FATA region – where previously 'Mujahedeen' (or Holy Warriors) were trained to fight against the Soviet Union in Afghanistan between 1979 and 1989 – soon became a breeding ground for local and transnational militant groups. Due to the concentration of militants, and with the exception of a few strikes outside the region, the former-FATA region remained the target of almost all drone strikes

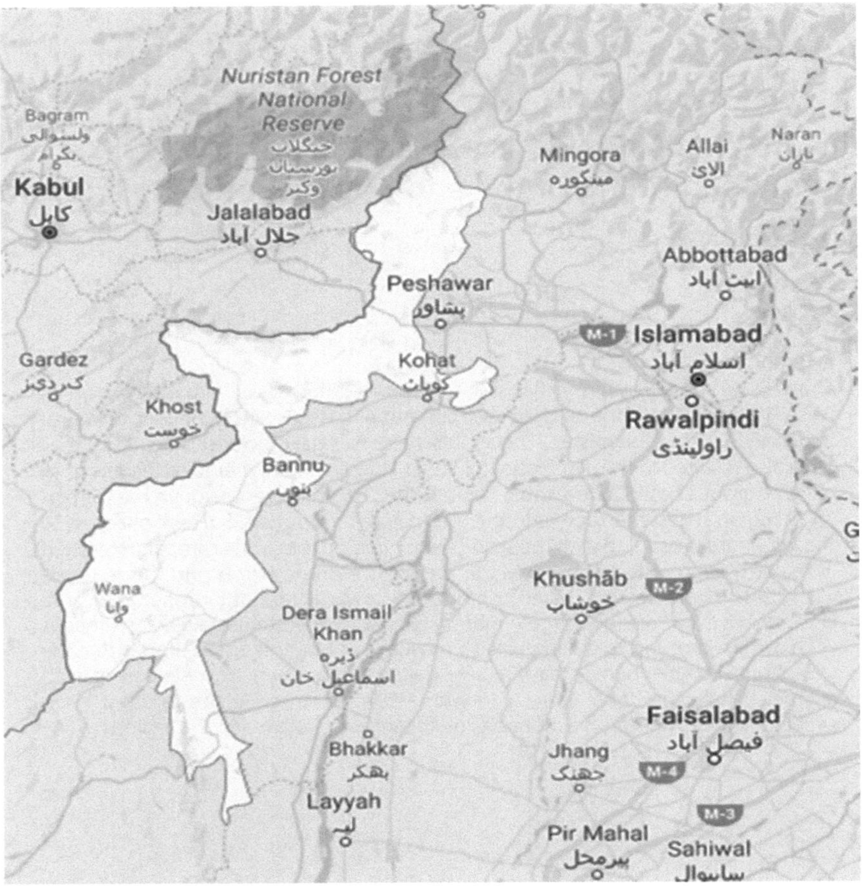

Figure 1. Pakistan's Pashtun 'tribal' region (in white) bordering Afghanistan.
Google, "Federally Administered Tribal Area"

conducted on the Pakistani soil by the U.S. With its first strike in 2004, the U.S. drone campaign in Pakistan, until the last strike in July 2018, comprised of over 400 strikes, mainly targeting various militant groups and leaders based in FATA's 'tribal agencies.'[1]

Due to a substantial decline in the drone strikes during the final months of the Obama administration (2016–2017), the U.S. seemingly excluded its drone operations from its counter-terrorism policy on Pakistan.[2] Even though eight drone strikes in Trump's first year of his presidency negated such perceptions, with no strikes in the first six months of 2019, the U.S. drone campaign in Pakistan may have finally come to an end. This apparent end has culminated out of different factors: President Trump's intention of a complete withdrawal from Afghanistan; the Pakistani military's counter-terror operations resulting in most of the terrorist groups moving into neighbouring Afghanistan; and, finally, a drastic decline in violence and violence-related casualties in Pakistan.

In this regard, this article provides a brief overview of the U.S. drone strikes in Pakistan (2004–2018) and the complications associated with these strikes. It then discusses contrasting opinions and perceptions of the U.S. drone strikes in Pakistan and argues why, even with their apparent success and effectiveness, these strikes have fuelled anti-Americanism sentiments in the country. In doing so, I argue why successive U.S. administrations preferred countries like Pakistan, with weaker democracies, to conduct its drone campaign as the media and the public narrative is generally shaped by the ruling elite, particularly the military establishment.[3] In its final section, the article discusses the possible reasons for the change in U.S. drone policy in Pakistan and discusses whether or not these strikes will continue in the near or distant future.

This article uses figures from New America (previously New America Foundation) and Bureau of Investigative Journalism's (BIJ) drone databases, along with employing prior literature on the topic to provide a brief historical analysis of U.S. drone strikes in Pakistan. The article also utilizes interviews and opinions from both the U.S. and Pakistani security and defence experts, as well as several perception surveys – including one conducted by the author in 2014 – on U.S. drone strikes conducted in Pakistan to explain why, even with their effectiveness in eliminating major terrorist targets and a tacit approval by the Pakistani state, these strikes, even today, remain one of the most contentious issues in the country. Finally, the article aims to connect these historical complications with current geopolitical developments to argue whether or not, even with a hiatus in strikes, the U.S. may target the 'tribal' region – which has become part of Pakistan's mainland territory since May 2018 – with drone strikes.

U.S. drone strikes in Pakistan

The U.S. drone campaign in Afghanistan-Pakistan (Af-Pak) region was 'legally' enabled by the passage of the Authorization for Use of Military Force (AUMF) by the U.S. Senate on 14 September 2001. It gave then-President George W. Bush the authority to use discretionary powers against those responsible for the 9/11 attacks. This was followed by the initiation of Operation Enduring Freedom (OEF) leading to a direct U.S.-led coalition invasion of Afghanistan and, significantly, the deployment of armed Predator drones to the country. In November 2001 the first lethal attack by an armed drone was conducted. After drone operators spotted a number of 'Arabs' gathered in a three-storey building in Kabul, a Predator, with assistance from an F-18 Hornet, fired missiles on the compound, killing Abu Hafs Al Masri, also known as Muhammad Atef, who was Al Qaeda's third in command and Osama Bin Landen's close aide.[4] This apparent success gave the Bush administration enough impetus to formally start the U.S. drone campaign in Afghanistan.

Even though the U.S. government shifted much of its focus to the war in Iraq after 2003, the search for Osama Bin Laden and Ayman Al Zawahiri was still underway. In the same year, Al Qaeda released a video of Bin Laden, alongside Zawahiri, in what was presumed to be a mountainous region in Pakistan's former-FATA region. After getting a tip that local Taliban may be protecting Al Qaeda fighters and Zawahiri, the Pakistani army launched an operation on 16 March 2004. With the military incurring heavy losses (approximately one hundred soldiers were killed), the Pakistani government was forced to sign a 'peace accord' with the local Taliban, who were represented by an ex-Afghan War veteran and local tribesman Nek Muhammad Wazir. In what seemed like a retreat and surrender for the Pakistani army, the peace agreement, known as the Shakai Peace Accord, was signed in April 2004 between the Pakistani army and the Taliban to halt military operations against the Taliban in Waziristan.[5]

On 19 June 2004, it was reported that Muhammad Wazir had been killed by Pakistani security forces after intercepting his fifteen-minute interview with a foreign news channel and tracing his location. Because of the precision of the night strike that killed Wazir, questions were raised on who, or what, actually killed Wazir. A local witness was quoted saying that Wazir was killed by a low-flying drone and not by the Pakistani army. This account by locals and the precision of the guided-missile that hit Wazir confirmed the first American drone strike on Pakistani soil. Since this first attack, the U.S. has conducted over 400 strikes in Pakistan, eliminating dozens of low, medium and top-level leaders belonging to Al Qaeda, Taliban, Haqqani Network, Pakistani Taliban and Islamic Movement of Uzbekistan (IMU), among others.[6]

President Obama's first term as the U.S. president saw a drastic spike in drone strikes, especially with his authorization of the infamous 'signature strikes,' whereby the targets were not based on hard evidence but rather on their behaviour and appearance as 'military-age' males who resembled terrorists.[7] Such signature strikes were also employed in former-FATA to target militants. Among these targets, Ayman Al-Zawahiri, the then number two of Osama bin Laden, was targeted twice; both attempts not only failed to kill him but also resulted in over 100 casualties. In one of the strikes in the village of Damadola (Bajaur Agency) in 2006, over 80 people, mostly students at a religious seminary, were killed. Therefore, questions were raised on the precision and effectiveness of drone strikes as causalities in the Damadola attack contradicted U.S. claims of precision and 'low collateral damage'.

Peter Bergen and Emily Schneider reported in 2014 that with the U.S. focused on Yemen and Somalia and the lack of infrastructure to oversee the drone operations, the Obama administration had effectively wrapped up its drone campaign in the Af-Pak region.[8] However, periodic strikes since 2014, even though less frequent than before, on important targets in Pakistan's 'tribal' areas and elsewhere suggested otherwise. In

May 2016, Mullah Akhtar Mansur, the successor to the deceased Taliban head Mullah Omar, was killed by a U.S. drone strike while travelling in Pakistan's Balochistan province.[9] With only nine strikes since Mansur was killed – the most most recent of which took place in July 2018 – it would seem the U.S. drone campaign has withered in Pakistan.

Conflicting figures, polarising debates

In over fifteen years of drone strikes in former-FATA, a number of top militant and terrorist leaders have been eliminated.[10] However, due to the secrecy surrounding the drone campaign and lack of access to the 'tribal' areas until recently, conflicting reports have emerged on the total number of strikes and the civilian loss of life in these strikes. Both the U.S. and the Pakistani governments, along with different human rights groups, have presented contrasting and conflicting drone casualty figures. The total number of drone strikes according to the Bureau of Investigative Journalism (BIJ)[11] stood at 430 by January 2018, whereas New America (NA)[12] quoted a total figure of 412. In terms of civilian deaths, there is also a discrepancy in numbers recorded in the BIJ and NA databases, especially as the NA database classifies victims between 'terrorists, civilians and unknown,' thereby understating the number of civilian casualties compared to the BIJ database. The NA database is predominantly focused on militant casualties and how drone strikes have negatively affected the operations of these militant organizations. On the other hand, the BIJ database focuses only on highlighting the number of casualties, mainly civilians. Therefore, different aims have led to different outcomes in terms of reporting numbers and effectiveness of these strikes.

When it comes to official state figures, according to a statement submitted by Pakistan's Ministry of Defence in 2013 to the National Senate (Upper House of the Parliament) there were no civilian casualties resulted from drone strikes between 2012 and 2013.[13] However, this official claim was contradicted in the aforementioned databases, both of which reported four civilian casualties in 2013. Additionally, figures released by Pakistan's defence ministry in 2013 also claimed that 67 civilians were killed in all drone strikes between 2004 and 2013, which was a figure much lower than that in the BIJ and NA databases.[14] In his final year of presidency, Obama made the drone civilian casualty report public. The civilian casualty rates were, however, again underestimated between 2.6 per cent to 4.3 per cent.[15]

A number of reasons led to such discrepancies in drone casualty figures. The three major contributing factors, in this regard, were: misidentification of drone targets; battle damage assessment (BDA) through air surveillance; and the irregular and asymmetric nature of the enemy.[16] Because BDA is mainly conducted through air surveillance, the irregular nature of the enemy targets coupled with the 'possibility of misidentification' makes it difficult to assess the civilian death toll accurately.[17] Additionally, Micah Zenko believes that the U.S. administrations' unrealistic trust on the figures provided by intelligence agencies overseeing the drone operations also led to an unrealistically low number of civilian causalities.[18] Such faith, leading to an underestimation of numbers, was witnessed when in 2011, the then-White House counterterrorism adviser, John Brennan claimed that there had not been a single incident of collateral damage in drone strikes. His statement was refuted by the data published by the BIJ that put the number of civilians killed during the period Brennan was referring to between 136 (minimum) and 339 (maximum).[19]

Finally, one of the major reasons for the unclear and ambiguous number on drone casualties is that many areas in former-FATA remained 'no-go zones' and have mostly administered by the Pakistani military, especially since 2002, due to various military operations. This made it nearly impossible for both local and foreign media personnel to corroborate the figures provided by both the U.S. and the Pakistani governments. The Pakistani military, especially since 2002, has controlled the security and counter-terror 'narrative' in the country. Various cable leaks and news stories suggest that Pakistani officials had a tacit understanding with successive U.S. administrations over drone strikes. Due to this understanding, the CIA was allowed to carry drone strikes in the former-FATA region, however, the official position of successive Pakistani governments, both civilian and military, was based on issuing formal condemnations for most of the strikes. More importantly, both the military and civilian leadership believed that drones carried 'some utility' in eliminating terrorists from the tribal region due to which an understanding was reached with the US to carry out these strikes.[20] This tacit understanding also explained the complex nature of Pakistan's postcolonial democracy, where the military, since the country's independence in 1947, has maintained a substantial hold over foreign and domestic security policy. Christine Fair, in a review article, succinctly describes these dynamics in the following words:

> The current guise of the (drone) program also undermines the development of civilian control over Pakistan's military. Pakistan's army and ISI seems to be perpetual beneficiaries of this program: the United States kills their enemies, while they take neither credit nor blame. At the same, Pakistani politicians who protest the drones but have no ability to stop them appear continuously ineffective to the Pakistani voter who largely loathes the drone program. This perceived incompetence in turn buttresses the army's claim that politicians are feckless and unable to protect Pakistan thus providing the continual justification for the army to dominate the portfolios of defense and national security.[21]

However, even with such socio-political complications in Pakistan, the U.S. drone strikes have eliminated a number of high-profile terrorist leaders in Pakistan – namely Nek Muhammad Wazir, Abu Haitham al Yemeni, Abu Hamza al Rabia, Baitullah Mehsud, Hakimullah Mehsud and Mullah Akhtar

Mansur, among others. However, these strikes have also remained a subject of strong contention. Hence, even with their success, understandably, drone strikes have also led to debates on their legality, legitimacy, and effectiveness.

In terms of legal justification, officials in Washington have often cited the AUMF for conducting counter-terror drone operations on foreign soil. The following passage in the AUMF suggests the nature of authority afforded to the President of the United States.

> That the President is authorized to use all necessary and appropriate force against those nations, organizations or persons he determines planned, authorized, harbored, committed, or aided in the planning or commission of the attacks against the United States that occurred on 11 September 2001, and to deter and pre-empt any future acts of terrorism or aggression against the United States.[22]

In addition to the AUMF, the National Defense Authorization Act of 2012, asking the U.S. president to take all possible measures to go after terrorists, both home and abroad, was also cited to argue for the legality of the drone operations. However, the United Nations Charter, Section 4 of Article 2, prohibits any member state from using force against another member state unless either the Security Council allows such an action or the host country agrees on foreign intervention. Strictly within this framework, the U.S. drone strikes can be deemed illegal. However, it was Pakistan's leaders – especially military dictator Pervez Musharraf (1999–2008) – whose tacit approval of drone strikes weakened Pakistan's legal case in international fora. Musharraf publicly admitted to permitting 'a few' drone strikes in the past, pointing towards an active intelligence-sharing agreement, albeit a tacit one, with the CIA.[23] Moreover, Wikileaks cables reveal that the Pakistani military asked for more drone strikes as well as drone assistance from the U.S.[24]

A 2016 report by Larry Lewis and Diane Vevrichek argued that even though U.S. drone strikes have remained effective in limiting the operational capabilities of terrorists in the region in the short run, they might have long term repercussions in terms of creating grievances among the local population, leading to greater radicalisation and militant recruitment.[25] Moreover, many anti-drone writers argue that these strikes only fan the flames of the anti-Americanism sentiment in FATA (and in Pakistan), in addition to providing justification to the militant narrative.[26] This sentiment, thereon, helps the militant groups to recruit youngsters as well as relatives of the affected families. One such example is of Faisal Shahzad – the so-called 'Times Square bomber' – who claimed his motivation (to commit terrorist activities) stemmed from the innocent lives that were lost in U.S. drone strikes. A militant organizer was also quoted saying that the drone strikes, the resultant loss of life, and widespread coverage on the media had made it easier for terrorist groups to recruit young fighters.[27]

However, Aqil Shah and Farhat Taj argue that claims of drone blowback are a myth.[28] Aqil Shah's detailed study, consisting of 167 interviews from 'well-informed adults from North Waziristan' and an official Pakistan police survey of 500 detained terrorists from the Sindh province, finds 'no evidence of impact of drone strikes on recruitment of terrorists.'[29] Farhat Taj has directly criticized U.S. media for propagating suspect figures related to the number of civilian casualties from drone strikes. According to her, neither the Pakistani civilian government nor anyone else had access to the 'tribal' region to verify these numbers. She further argued that the same numbers, presented by the American media, were used by Pakistan's right-wing political parties as well as terrorist groups to generate anti-US feelings.[30] In addition to that, Daniel Byman also argues that the anti-drone sentiments expressed by Pakistanis are in surveys conducted by anti-drone organisations 'resulting in biased samples,' whereas Pakistanis have historically held an unfavourable view of the U.S. even before the drone campaign expanded in the country.[31]

In its initial years of operations, the lethal U.S. drone campaign in FATA was severely criticised for the high civilian casualty rates. New America reports that in 2006, nearly 100 per cent of strikes in Pakistan resulted in civilian casualties. However, consistent criticism and calls for transparency helped in reducing these casualties – strikes in 2013 reported nearly zero civilian casualties. The New America report also claims that all drone strikes under former President Obama resulted in 3000 casualties, with terrorist leaders only making up two per cent of the total deaths.[32] Along with the high civilian death toll, drone strikes have also been criticised for using a 'double-tap' method – a drone strike following another strike where people rush to save targets of the previous strike.[33] Christof Heyns – former UN special rapporteur on extrajudicial killings (2010–2016) – even argued that such strikes could come under the definition of 'war crimes.'[34]

Conversely, advocates of drone strikes have made four arguments in their support for the program. For them, drone strikes: (1) caused minimum collateral damage; (2) were efficient in taking out high-value targets; (3) put pressure on terrorists; and (4) improved the cost-benefit ratio.[35] These arguments are supported by some locals from the 'tribal' areas, where a young student from North Waziristan, in 2013, told the author that the locals preferred drones over Pakistani military operations due to the former's precision and lower collateral damage.[36] Christine Fair makes a similar argument. According to her, drones were the best available alternative for the U.S. especially because the Pakistani army operations resulted in the internal displacement of locals and destroyed infrastructure.[37] Avery Plaw also believes that drones not only saved lives – both American and others – and weakened terrorist groups, but also minimised civilian casualties where such casualties could not be avoided.[38] Understandably, conflicting arguments on the U.S. drone strikes have also generated conflicting perceptions on the utility of such strikes in Pakistan's 'tribal' areas. These perceptions are discussed in detail in the forthcoming section.

Conflicting perceptions

Since the initiation of the U.S. drone campaign in former-FATA, there has remained widespread media, public and state opposition in Pakistan against these strikes. The disparity between the official positions taken by the U.S. and Pakistani governments on the legality and legitimacy of drone strikes, along with the semi-constitutional status of former-FATA before its merger in May 2018, has left a major grey area in international law on the question of 'foreign strikes' on Pakistani soil. Since the first drone strike in 2004, the Pakistani authorities and military, contrary to the U.S. claims, have 'officially' denied giving their consent to the U.S. for conducting drone strikes in FATA. This denial influenced the general discourse surrounding the drone strikes in Pakistan where the general public considered them a violation of the country's sovereignty.[39] Moreover, this policy of denial and ambiguity from the Pakistani side was actively pursued between 2003 and 2013 as, during this time, Pakistan witnessed a period of military dictatorship (1999--2008) and a weak democratic government (2008–2013).

Numerous surveys conducted in Pakistan among the population suggest a variety of opinions on U.S. drone strikes. A survey conducted by PEW research centre noted in 2013 that approximately 68 percent of Pakistanis opposed the U.S. drone strikes in former-FATA.[40] Some local accounts of drone strikes also suggested drone strikes created 'some' backlash among the local population. Pir Zubair Shah, a former-FATA resident, while writing on the issue of drone strikes, interviewed locals who told him that respected locals and 'tribal' elders, with no connections to militant groups, were killed in indiscriminate drone strikes.[41] The death of one such 'tribal' elder, Daud Khan, left thousands in his tribe and village with a feeling of animosity and hatred towards the U.S.[42] Abdul Basit, counter-terrorism researcher based in Singapore, argues that even though the U.S. drone strikes have remained 'tactically effective' in eliminating terrorists, they were 'strategically disastrous' as they created more militants given the collateral damage they caused.[43]

In terms of political opposition, Pakistan has also witnessed large scale protests against drone strikes, especially in the Khyber Pakhtunkhwa province. Imran Khan, formerly a prominent opposition figure but currently the country's prime minister, led a famous anti-drone protest in 2013 calling for the blockade of NATO supply lines in the country until drone strikes were halted.[44] Khan criticised the killing of Hakeemullah Mehsud, the then head of Tehrik e Taliban Pakistan (TTP), calling it a blow to the peace process in Pakistan. It is worth noting that not only was Mehsud the most wanted terrorist in the country at that time but was also responsible for one of the deadliest sprees of terrorism in the country. A protest against his death suggested why the drone narrative in Pakistan was not only 'confused,' but was also manufactured to create a sentiment of anti-Americanism in the country.[45] Similarly, in 2016, soon after the killing of Taliban head Akhtar Mansur, there were protests in both Khyber Pakhtunkhwa and Balochistan provinces, mainly by the right-wing parties and groups, condemning the U.S. and the killing of Mansur.[46]

Politicians like Imran Khan have also argued that the U.S. drone campaign helps militant organizations recruit because strikes radicalise the youth and are thus a major stimulant to terrorism.[47] There is, however, limited, or no, evidence to back such claims.[48] Experts and academics are of the view that the discourse on drones and the generalized opposition to it hold little substantial ground.[49]

Similarly, in 2010, a group of politicians and NGOs signed a declaration in Peshawar – known as the Peshawar Declaration – in support of the drone strikes. However, the signatories and participants later faced harassment from Pakistan's intelligence agencies as the declaration also criticised the Pakistani army's role in the escalation of conflict in the 'tribal' areas.[50] Aqil Shah, while mentioning various accounts of former-FATA residents, also argues that locals in the region believed that drone strikes were more precise than Pakistani military operations, which caused far more collateral damage.[51] It was due to this collateral damage that 'tribal' Pashtuns expressed resentment towards the military operations, which, along with the damage, also resulted in human rights violations, including enforced disappearances, in the region.[52]

In terms of media bias and narrative framing on U.S. drone strikes in Pakistan, a study by Christine Fair and Ali Hamza concluded that the majority of Pakistanis were unaware of the issue of drone strikes and, therefore, their opinions were framed by the major media networks that already carried a negative bias towards the U.S.[53] Sarmad Tariq, a Pakistani writer, wrote that his friend's aunt was killed by the indiscriminate firing of the Pakistani military helicopters in the 'tribal' areas, whereas according to his conversations with the locals, U.S. drone strikes were much precise resulting in lower collateral damage.[54] Advocating the support for drones, a reader, in his letter to the editor of Pakistan's renowned English daily *Dawn* argued:

> The media too, inadvertently, follows the line of the religio-political parties and creates hype and makes it look as if the Americans have done great harm to Pakistan while the other set of foreigners, i.e. Arab, Chechen, Uzbek militants, have played no role in a persistent effort to destabilise Pakistan.[55]

Much of the anti-drone discourse can also be linked to sentiments of anti-Americanism in Pakistan, mostly perpetuated through, and by, the mainstream media.[56] Moreover, due to limited access to the former-FATA region, and with no access to some parts even today, very few sources could quote or report on the exact number of casualties caused as a result of these strikes. This also leaves a big discrepancy in the figures, as discussed above, presented by advocacy centres, governments and human rights watchdogs.

It is also argued that due to a change in strategy of CIA operations, the number of civilian casualties may have dropped to as low as 3 percent, though reporting in media and the unclear distinction of militants and civilians has created an unfavourable discourse among the public.[57] However, Pakistan's formal condemnation of these strikes and shaping of anti-US media discourse – even while allegedly having a tacit agreement with the CIA and the U.S. government – and strong opposition of national political leaders generated a generally negative public sentiment against the U.S. and the drone strikes in the country.[58]

When it comes to the drone strike perception within the 'tribal' areas, it is somewhat mixed. A survey conducted by Community Appraisal and Motivation Programme (CAMP) in the former-FATA region noted that in 2011 only 3 percent of the respondents believed that the U.S. drone strikes were a major issue faced by the region.[59] The same survey, which questioned respondents from all 'agencies' (tribal units) of former-FATA, also reported that a majority of the people considered the lack of law and order as the major problem faced by the region, and chose terrorism as the biggest threat to their security. However, even with these opinions, the majority of respondents believed that the U.S. drone strikes were 'not justified.' Similarly, a survey conducted by the *Economist* in 2013 suggested that locals in former-FATA supported drone strikes over Pakistani military operations due to their precision.[60] This perception was supported by a resident of North Waziristan, who, while talking to *ABC News* in 2008, said that 'the attacks have become precise. In a village, if they (the U.S.) want to hit a house in the middle of the village and it is surrounded by other houses, the missile would come and hit that one house only.'[61] Farhat Taj also argues that people in the 'tribal' areas mostly carried on with their business when they heard a drone hovering over their heads (believing them be more precise), but ran for shelter when the saw Pakistani air force jets.[62]

Perception survey in FATA (2014)

It was reflected from the discussion above that a significant number of Pakistanis, outside the 'tribal' areas, opposed drone strikes conducted by the U.S. However, opinions within the 'tribal' areas indicated a relatively different set of feelings towards drones and military operations. Similar feelings were also reflected in a survey conducted by the author in 2014 among the residents of the former-FATA region.[63] In this survey, 109 out the 132 total respondents aged between 18 and 39, with 66 percent of the respondents having 'some' educational background. The survey focused on the overall perception and effectiveness of the U.S. drone strikes among the residents of the former-FATA region. Majority of the respondents in the survey, around thirty-four per cent, believed that terrorism – and not drones – was a major threat faced by the 'tribal' region (see Figure 2 below). Around one percent of the respondents believed that drone strikes were a major issue faced by the region.

However, as discussed above, even with this lack of importance attributed to drone strikes, 69 percent of the respondents did not support the U.S. drone strikes in FATA. Interestingly, around 26 percent of the respondents, a lower but significant number, support the drone strikes in the 'tribal' areas (see Figure 3 below).

The respondents who supported these strikes believed that drones were precise and resulted in lower collateral damage. Moreover, even with a majority of the respondents against these strikes, 59 percent of the respondents believed that Pakistan did have a tacit drone agreement with the United States, allowing the CIA to conduct these strikes on its territory.

Has the US finally ended its drone campaign in Pakistan?

The gradual halt in the U.S. drone strikes in Pakistan represents a stark contrast in President Trump's counter-terrorism policy as similar strikes have escalated both in Yemen and Somalia since he assumed power. Even

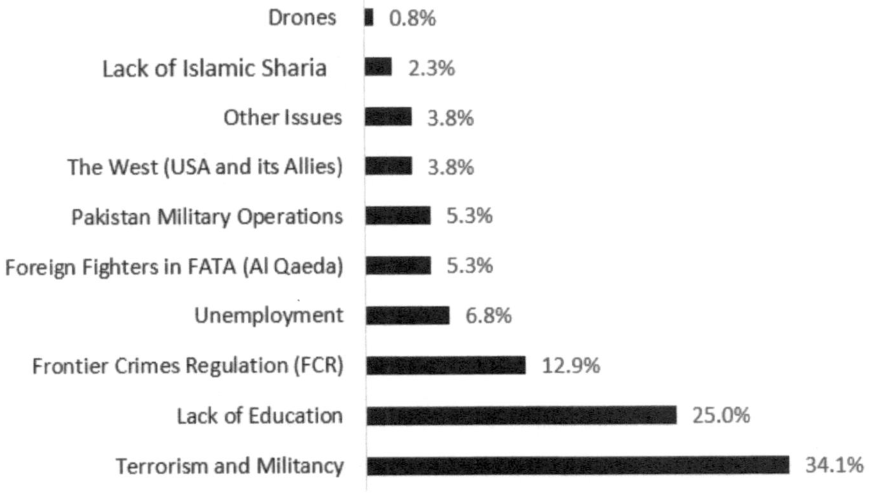

Figure 2. Public perception in former-FATA on major issues in the region.
Yousaf, "CIA Drone Strikes in Pakistan."

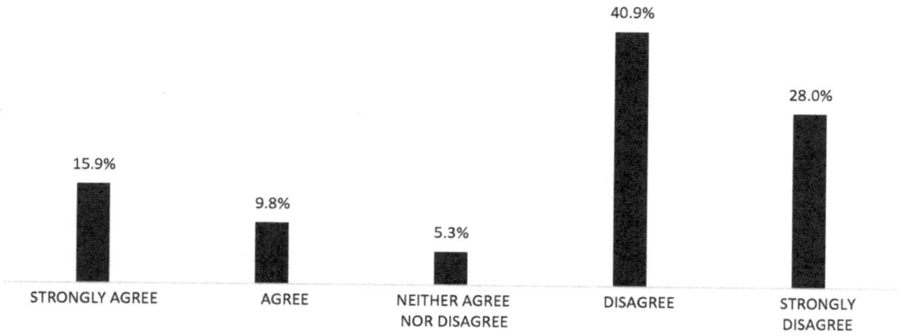

Figure 3. Support for U.S. drone strikes in Pakistan's former-FATA region.
Yousaf, "CIA Drone Strikes in Pakistan."

though this trend might be confusing for many security experts, a number of geopolitical changes in Afghanistan and Pakistan give some indication of this halt. These changes – as well as a change in direction of the U.S. security policy vis-à-vis Afghanistan – are echoed in the opinions of security, defence and foreign policy experts both in Pakistan and the U.S.

The think tank analyst Michael Kugelman believes that the U.S. drone campaign in Pakistan has 'likely run its course for several reasons.'[64] The first reason, according to Kugelman, is that the U.S. has eliminated most of its desired 'Pakistan-based targets.' Secondly, the U.S.-Pakistan relationship has deteriorated to the point that 'Washington is unlikely to get the intelligence assistance and tacit support from Pakistan that enables the U.S. to carry out drone strikes.' 'To be sure, there could be times when the U.S. identifies a target by itself and decides to act unilaterally. But for the foreseeable future, I wouldn't expect many drones over Pakistani skies,' argues Kugelman.

On the other hand, for security experts in Islamabad, one of the primary reasons for the decline, and a gradual halt, in the U.S. drone strikes is the excessive Pakistani military footprint in the Pashtun 'tribal' areas. Imtiaz Gul, the executive director for the Center for Research and Security Studies (CRSS) in Islamabad, believes that successive operations by the Pakistani army, pushing militants out of the 'tribal' areas is one of the major reasons for the drone strike halt in Pakistan.[65] According to Gul, 'the first major change or indicator that has apparently led to the U.S. reduction in drone strikes, especially since 2015, is major military operations, especially Zarb e Azb (or Sharp Strike), conducted by the Pakistani army in the "tribal" areas.' Gul also believes that these operations have resulted in a drastic decline in violence and violence-related casualties. Between 2010 and 2015, the average incidents of yearly violence stood at 1753, average civilian casualties 2168 and average military casualties 616. However, since 2015, when major military operations started, until the conclusion of 2018, the average yearly incidents of violence stood at 482, average civilian casualties 552, and average military casualties 250.[66] This significant decline in violence and casualties is a result of military operations resulting in not only the elimination of terrorist groups and networks but also forced these groups to move to neighbouring Afghanistan.

In terms of future of U.S. drone strikes in former-FATA, Gul believes that even though drones are 'cheap, non-risky and cost-effective in terms of hounding, eliminating and pushing the terrorists out of the "tribal" areas,' the Trump administration would hesitate in launching such strikes due to two major reasons. The first reason is FATA's integration with 'mainstream' Pakistan and unlike in the past, where the 'tribal' areas and their administration fell outside the scope of the Pakistani constitution, any future strikes would raise serious questions of violation of Pakistan's sovereignty.[67] This is because any strike in the 'tribal' areas (now known as 'tribal' districts) would be considered a direct strike on Pakistan's sovereignty. According to Gul, the second reason why he does not see any future drone strikes, at least in the near future, is because such strikes might risk the Pakistani cooperation with the U.S. on Afghan peace talks with the Taliban. Pakistan is reportedly facilitating the Taliban peace talks with the U.S. administration. In October 2018, Pakistani authorities agreed to release Mullah Abdul Ghani Baradar, a senior Taliban leader in Pakistan's captivity, at the request of U.S. Afghan peace envoy Zalmay Khalilzad to facilitate the dialogue. Gul believes that such gestures and U.S. strikes cannot go hand in hand if Washington desires Pakistan's cooperation in the Afghan reconciliation process.

Abdul Basit, a counter-terrorism researcher based in Singapore, believes that even with their 'tactical' effectiveness in eliminating terrorists, he does not see any drone strikes happening in the 'tribal' areas in the near future.[68] This, he argues, is because the 'theatre of conflict' has shifted from the former-FATA region to Pakistan's Balochistan province, especially with recent attacks on Chinese installations in the province. Moreover, he also argues that major militant groups – mainly Haqqanis, TTP and Jamaat ul Ahrar – have now moved into neighbouring Afghanistan, making it less likely for the U.S. to conduct any drone strikes in Pakistan's Pashtun 'tribal' region. Basit, similar to Gul, also argues that the U.S. intentions of a complete withdrawal from Afghanistan reduce the likelihood of any future drone strikes in Pakistan.

The narrative within the Pakistani military is also somewhat similar to that of civilian security experts. A senior retired military officer believes that even though the previous 'agreements' on drone strikes were productive, there is little possibility of such agreements in the present government setup. According to him, the current PTI government, led by Prime Minister Khan, and the military establishment are on the same page in terms of national security and foreign policy.[69] Therefore, it would be difficult for the military to have a tacit understanding similar to that of the Musharraf-era with the CIA without taking the prime minister on board. The military also understands that Prime Minister Khan has remained one of the most vocal critics of the U.S. drone strikes, among all national political parties, and therefore any future strikes, with the military's tacit consent, could create friction between the civilian and military leadership. Pakistan's decision to fence the Afghan border as means of reducing cross-border movement of terrorists might also have played a major part in the US decision to reduce drone strikes. Even though Afghanistan still contests the legality of the border – commonly known as the Durand Line – and stakes its claim over parts of northwest Pakistan, the Pakistani military has gone ahead with erecting a fence in order to display its commitment to stopping cross-border terrorist movement.

In the past, President Obama's motivation for conducting drone strikes on the Taliban, Al Qaeda and the Haqqani network was rooted in protecting the large contingent of coalition soldiers based in Afghanistan. However, a limited number of soldiers now remain in Afghanistan. Moreover, most of the drone strikes in Pakistan targeted either the Haqqani network or other Afghan Taliban factions; all of whom were a direct threat to the coalition forces present in Afghanistan. With these groups now mostly eliminated from the 'tribal' areas and pushed into neighbouring Afghanistan, the chances of any strikes in the near future remain low. However, one cannot rule out of the possibility of a 'surprise strike' as some traces of militancy and terrorism still remain in the region, especially the Maulvi Nazir group, alleged by the U.S. and Afghanistan to be receiving clandestine support from the Pakistani military.

David Sterman – senior policy analyst at New America – also believes that the halt in U.S. drone strikes in Pakistan is mainly due to a combination of different factors. He argues:

> By our tracking, it is just a few weeks short of a full year without a reported (drone) strike in Pakistan and what is more, there was an almost five-month pause broken by a single strike before that, and a long pause as well under the Obama administration and early Trump administration. With the number of troops in Afghanistan having wound down from its peak (also the peak of strikes in Pakistan) and Pakistan's own military efforts in (former) FATA, there is far less need for strikes and the (Taliban) negotiations may also be shaping the decision to hold back; though the pauses seem to be driven by something larger and more long term.[70]

The aforementioned discussion suggests that the U.S. would not hesitate in launching another strike if the presence of a high-profile terrorist target is confirmed in the 'tribal' region or Balochistan province. This argument can be backed by the targeting for Mullah Akhtar Mansur, as the strike that targeted him came at a time when the drone strikes were at an all-time low in Pakistan. President Trump's recent revocation of the Obama-era's 2016 policy of reporting drone figures also suggests that the U.S. president would not hesitate in launching a strike on Pakistani soil, especially the 'tribal' region if credible intelligence is put before him on a high-value target. Such a strike can especially become highly probable if the U.S. perceives that Pakistan is not 'cooperating' in the Afghan peace process. This is because many in Washington and Kabul still believe that Pakistan holds a significant 'sway' over the Taliban; a claim that defence and security experts in Islamabad now refute. However, it can be concluded that with the ongoing U.S.-Taliban talks in Afghanistan and Pakistan's perceived 'facilitative' role in the talks, the U.S. administration would not risk launching a strike on Pakistan and irk the civil-military leadership in Islamabad.

Conclusion

The U.S. drone strikes in Pakistan's former-FATA region have remained one of the most complex issues that Pakistan has faced since the global war on terror began in 2001. Due to the high influx of Al Qaeda and Taliban fighters in Pakistan's Pashtun 'tribal' belt, successive U.S. administrations used drones as a cost-effective precision tool, mainly because it helped in circumventing certain international legal obligations. However, as discussed in detail in this article, even with their effectiveness, the U.S. drone strikes have resulted in polarising debates on their legality, effectiveness and legitimacy. Much of the negative discourse surrounding these strikes in Pakistan can mainly be attributed to the Pakistani military's control over the security and media narrative in the country. However, after years of drone campaign and over 400 strikes, recent geopolitical developments both in Afghanistan and Pakistan suggest that the probability of future strikes in Pakistan is very low. This is because not only is the U.S., under President Trump, working towards a complete withdrawal from Afghanistan, but this withdrawal also requires Pakistan's 'facilitation' in terms of 'peace talks' with the Taliban. Moreover, a heavy Pakistani military footprint in the 'tribal' areas and successive military operations has pushed most of the militant groups into neighbouring Afghanistan. These factors make it less likely for the U.S. to conduct any drone strikes in the near future in Pakistan. However, as argued in the final section of this article, if provided with actionable intelligence on high-profile militant and terrorist targets in the former-FATA region, the U.S. President would not hesitate in ordering a strike on Pakistani soil.

Notes

1. Before FATA's integration with the Khyber Pakhtunkhwa province, it consisted of 7 'tribal agencies' and 6 'Frontier Regions.' These agencies and districts, after

the region's integration with the Khyber Pakhtunkhwa province in May 2018 after a constitutional amendment, are now known as 'Tribal' Districts.
2. Bergen and Schneider, "Obama has put Pakistan drone war on hold."
3. For detailed discussion on Pakistan's weak 'postcolonial' democracy, see Yousaf, "Pakistan's Colonial Legacy."
4. Grey, "U.S. Kills Al Qaeda Leaders by Remote Control."
5. Kilcullen, "Terrain Tribes and Terrorists."
6. New America, "Drone Strikes: Pakistan."
7. Ackerman, "US to Continue 'Signature Strikes'."
8. See note 2 above.
9. See note 7 above.
10. Ibid.
11. TBIJ, "Strikes in Pakistan."
12. See note 7 above.
13. Ansari, "Analysis: Drone Discrepancy."
14. Associated Press, "Pakistan Criticised over Drone Victim Numbers."
15. Grossman, "Trump Cancels Drone Strike Civilian Casualty Report."
16. Lewis, "Drone Strikes in Pakistan," 17.
17. Ibid.
18. Zenko, "Do Not Believe the U.S."
19. Singh, "Drone Strikes Kill Innocent People."
20. Amnesty-International, "Will I Be Next?," 52–53.
21. Fair, C. Christine. "Drones, spies, terrorists, and second-class citizenship in Pakistan.
22. Grimmett, "Authorization for Use of Military Force," 6.
23. Ahmed, "Rethinking Anti-Drone Legal Strategies."
24. Ibid.
25. Lewis and Vavrichek, "Rethinking the Drone War."
26. Hijazi et al., "Psychological Dimensions of Drone Warfare"; and Afzal, "On Pakistani Anti-Americanism."
27. Adams and Nasir, "Inside the Mind of the Times Square Bomber."
28. Shah, "Drone Blowback in Pakistan; Ibid, "Do U.S. Drone Strikes Cause Blowback"; and Taj, "Drone Attacks."
29. Shah, "Do U.S. Drone Strikes Cause Blowback," 48–49.
30. Taj, "Drone Attacks."
31. Byman, "Why drones work."
32. Slaughter, "Celebrating 20 Years of New America."
33. Williams, "New Light on Cia 'Double Tap'"; and Melancon, "What's Wrong with Drones."
34. Williams, "New Light on Cia 'Double Tap'".
35. Boyle, "The Costs and Consequences of Drone Warfare."
36. Yousaf, "Drone Strikes."
37. Fair, "For Now, Drones Are the Best Option."
38. Plaw, "Drone Strikes Save Lives, American and Other."
39. Peralta, "In Pakistan, Thousands Protest against U.S. Drone Strikes."
40. Drake, "Report Questions Drone Use."
41. Shah, "Analysis: US Drone Strikes in Pakistan's Tribal Areas Create Backlash."
42. Ibid.
43. Author's personal communication (interview) with Abdul Basit through email (5 June 2019).

44. Waraich, "Cricketer Turned Politician Imran Khan."
45. Fair and Hamza, "From Elite Consumption to Popular Opinion."
46. Bureau-Report, "Protest Rallies Held in Kp."
47. Dawn, "Imran Criticises US Drone Policy."
48. See note 27 above.
49. Fair, Kaltenthaler and Miller, "Pakistani Opposition to American Drone Strikes."
50. Economist, "Drop the Pilot."
51. Shah, "Do U.S. Drone Strikes Cause Blowback," 58.
52. Shafiq, "The War on Terror and the Enforced Disappearances in Pakistan."
53. See note 46 above.
54. Tariq, "Drone Strikes Kill Militants."
55. Bokhari, "Misplaced Sense of Sovereignty."
56. Afzal, "On Pakistani Anti-Americanism."
57. Plaw et al., "Practice Makes Perfect."
58. Fair, Kaltenthaler and Miller, "Pakistani Opposition to American Drone Strikes"; and Waraich, "Cricketer Turned Politician Imran Khan."
59. Shinwari, "Understanding Fata."
60. See note 51 above.
61. Schifrin, "Us Drone Strikes with Deadly Accuracy."
62. See note 31 above.
63. Yousaf, "CIA Drone Strikes in Pakistan."
64. Author's personal communication with Michael Kugelman through email (12 June 2019).
65. Author's personal communication with Imtiaz Gul through email (30 May 2019).
66. South Asia Terrorism Portal, "Fatalities in Terrorist Violence in Pakistan 2003–2019, South asia terrorism portal."
67. Yousaf, Rashid, and Gul, "Fata Tribes: Finally out of Colonial Clutches."
68. Author's personal communication with Abdul Basit through email (5 June 2019).
69. A high-ranking retired military officer, also previously actively involved in Pakistan's counter-terror operations, shared these views with the author through email, on condition of anonymity (29 May 2019).
70. Author's personal communication with David Sterman through email (10 June 2019).

Disclosure statement

No potential conflict of interest was reported by the author.

ORCID

Farooq Yousaf http://orcid.org/0000-0003-1566-8161

Bibliography

Ackerman, Spencer. 2017. "Us to Continue 'Signature Strikes' on People Suspected of Terrorist Links." *Guardian*, September. Accesed May 20, 2019. https://www.theguardian.com/us-news/2016/jul/01/obama-continue-signature-strikes-drones-civilian-deaths

Adams, Lorraine, and Ayesha Nasir. 2010. "Inside the Mind of the Times Square Bomber." *Guardian*, September. Accessed May 20, 2019. https://www.theguardian.com/world/2010/sep/19/times-square-bomber

Afzal, Madiha. 2013. "On Pakistani Anti-Americanism." *Brookings*, November. Accessed April 25, 2019. http://www.brookings.edu/research/opinions/2013/11/14-pakistani-anti-americanism-afzal

Ahmed, Dawood I. "Rethinking Anti-Drone Legal Strategies: Questioning Pakistani and Yemeni Consent." *Yale J. Int'l Aff.* 8 (2013): 1–11.

Amnesty-International. *"Will I Be Next?" Us Drone Strikes In Pakistan*. London, UK: Amnesty International Publications, 2013.

Ansari, Neha. 2013. "Analysis: Drone Discrepancy." *The Express Tribune*, November. Accessed June 2, 2019. https://tribune.com.pk/story/625508/analysis-drone-discrepancy

AP. 2013. "Pakistan Criticised over Drone Victim Numbers." *Al Jazeera*, October. Accessed June 1, 2019. https://www.aljazeera.com/news/asia/2013/10/pakistan-lashed-low-drone-victim-numbers-201310316713239576.html

Bergen, Peter, and Emily. Schneider 2014. "Obama Has Put Pakistan Drone War on Hold", CNN, May. Accessed November 30, 2019. https://edition.cnn.com/2014/05/23/opinion/bergen-obama-terrorism-pakistan-drones/index.html

Bokhari, S.G.B.S. 2008. "Misplaced Sense of Sovereignty." *Dawn News*, September. Accessed May 11, 2019. https://www.dawn.com/news/874450

Boyle, Michael J. "The Costs and Consequences of Drone Warfare." *International Affairs* 89, no. 1 (2013): 1–29. doi:10.1111/1468-2346.12002.

Bureau-Report. 2016. "Protest Rallies Held in Kp, Fata against Drone Attacks." *The News*, June. Accessed May 26, 2019. https://www.thenews.com.pk/print/126952-Protest-rallies-held-in-KP-Fata-against-drone-attacks+TheNewsInternational-Peshawar

Byman, Daniel. "Why Drones Work: The Case for Washington's Weapon of Choice." *Foreign Aff* 92 (2013): 32–43.

Dawn. 2013. "Imran Criticises Us Drone Policy." *Dawn News*, January. Accessed May 26, 2019. https://www.dawn.com/news/780563/imran-criticises-us-drone-policy

Drake, Bruce. 2013. "Report Questions Drone Use, Widely Unpopular Globally, but Not in the U.S." *Pew Research Center*, October. Accessed April 11, 2019. http://www.pewresearch.org/fact-tank/2013/10/23/report-questions-drone-use-widely-unpopular-globally-but-not-in-the-u-s/

The Economist. 2013. "Drop the Pilot." *The Economist*, October. Accessed April 15, 2019. http://www.economist.com/news/asia/21588142-surprising-number-pakistanis-are-favour-drone-strikes-drop-pilot

Fair, C. Christine. 2013. "For Now, Drones are the Best Option." *The New York Times*, September. Accessed June 10, 2019. https://www.nytimes.com/roomfordebate/

2012/09/25/do-drone-attacks-do-more-harm-than-good/for-now-drones-are-the-best-option

Fair, C. Christine. "Drones, Spies, Terrorists, and Second-class Citizenship in Pakistan." *Small Wars & Insurgencies* 25, no. 1 (2014): 205–235. doi:10.1080/09592318.2014.894061.

Fair, Christine, and Ali Hamza. "From Elite Consumption to Popular Opinion: Framing of the Us Drone Program in Pakistani Newspapers." *Small Wars & Insurgencies* 27, no. 4 (2016): 578–607. doi:10.1080/09592318.2016.1189491.

Fair, Christine, Karl Kaltenthaler, and William J Miller. "Pakistani Opposition to American Drone Strikes." *Political Science Quarterly* 131, no. 2 (2016): 387–419. doi:10.1002/polq.12474.

Google. (n.d.). *Federally Administered Tribal Area*. https://www.google.com/maps/place/Federally+Administered+Tribal+Area,+Pakistan/@33.0045693,68.3079839,7z/data=!3m1!4b1!4m5!3m4!1s0x38d7f648db7c706d:0x92d2f201433a32f5!8m2!3d32.667476!4d69.8597406

Grey, Stephen. 2001. "U.S. Kills Al Qaeda Leaders by Remote Control". *Fox News*, November. Accessed May 18, 2019. https://www.foxnews.com/story/u-s-kills-al-qaeda-leaders-by-remote-control

Grimmett, Richard F. 2007. *Authorization for Use of Military Force in Response to the 9/11 Attacks*. Vol. 16. CRS Report for Congress, (PL 107-40): Legislative History, Order Code RS22357.

Grossman, Nicholas. 2019. "Trump Cancels Drone Strike Civilian Casualty Report: Does It Matter?" *War on the Rocks*, April. Accessed June 2, 2019. https://warontherocks.com/2019/04/trump-cancels-drone-strike-civilian-casualty-report-does-it-matter/

Harris, Kathryn Elizabeth. 2016. "Asymmetric Strategies and Asymmetric Threats: A Structural-Realist Critique of Drone Strikes in Pakistan, 2004–2014." *Dissertation Virginia Tech*. Accessed June 28, 2019. https://vtechworks.lib.vt.edu/bitstream/handle/10919/64516/Harris_KE_T_2016.pdf?sequence=1&isAllowed=y

Hijazi, Alaa, Christopher J Ferguson, Harold Hall, Mark Hovee, F Richard Ferraro, and Sherrie Wilcox. "Psychological Dimensions of Drone Warfare." *Current Psychology* 38, no.5 (2017): 1285–1296.

Kilcullen, David J. 2009. "Terrain Tribes and Terrorists: Pakistan, 2006–2008." *Brookings Counterinsurgency and Pakistan Paper Series, no. 3*. Accessed June 16, 2019. https://www.brookings.edu/wp-content/uploads/2016/06/06_counterinsurgency_kilcullen.pdf

Lewis, Larry. 2014. "Drone Strikes in Pakistan: Reasons to Assess Civilian Casualties." *CNA - Centre for Naval Analyses*. Accessed May 10, 2019. https://www.cna.org/CNA_files/PDF/COP-2014-U-007345-Final.pdf

Lewis, L.L., and D.M. Vavrichek. *Rethinking the Drone War: National Security, Legitimacy, and Civilian Casualties in U.S. Counterterrorism Operations*. MCUP: CNA and Marine Corps University Press, 2016.

Melancon, Andree-Anne. "What's Wrong with Drines? Automization and Target Selection." *Small Wars and Insurgencies* 31 (2020): 3.

New-America. 2018. "Drone Strikes: Pakistan." *New America*, July. Accessed June 2, 2019. https://www.newamerica.org/in-depth/americas-counterterrorism-wars/pakistan/

Peralta, Eyder. 2013. "In Pakistan, Thousands Protest against U.S. Drone Strikes." *NPR*, November. Accessed April 10, 2019. http://www.npr.org/blogs/thetwo-way/2013/11/23/246887028/in-pakistan-thousands-protest-against-u-s-drone-strikes

Plaw, A. 2012. "Drone Strikes Save Lives, American and Other." *The New York Times*, November. Accessed June 8, 2019. https://www.nytimes.com/roomfordebate/

2012/09/25/do-drone-attacks-do-more-harm-than-good/drone-strikes-save-lives-american-and-other

Plaw, A., M.S. Fricker, and B.G. Williams. "Practice Makes Perfect?: The Changing Civilian Toll of Cia Drone Strikes in Pakistan." *Perspectives on Terrorism* 5 (2011): 5–6.

Schifrin, Nick. 2008. "Us Drone Strikes with Deadly Accuracy." *ABC News*. Accessed May 11, 2019. http://abcnews.go.com/International/Story?id=6289748

Shafiq, Aysha. "The War on Terror and the Enforced Disappearances in Pakistan." *Human Rights Review* 14, no. 4 (2013): 387–404. doi:10.1007/s12142-013-0282-2.

Shah, Aqil. 2016. "Drone Blowback in Pakistan Is a Myth. Here's Why." *The Washington Post*. Accessed May 12, 2019. https://www.washingtonpost.com/news/monkey-cage/wp/2016/05/17/drone-blow-back-in-pakistan-is-a-myth-heres-why/?utm_term=.bd33b6bd883e

Shah, Aqil. "Do US Drone Strikes Cause Blowback? Evidence from Pakistan and Beyond." *International Security* 42, no. 4 (2018): 47–84. doi:10.1162/isec_a_00312.

Shah, Pir Zubair. 2012. "Analysis: Us Drone Strikes in Pakistan's Tribal Areas Create Backlash." *PRI*, October. Accessed June 16, 2019. https://www.pri.org/stories/2012-10-10/analysis-us-drone-strikes-pakistans-tribal-areas-create-backlash

Shaikha, Shahzeb. 2013. "Drone Em All." *The Express Tribune*. Accessed May 28, 2019. https://tribune.com.pk/story/636911/drone-em-all/

Shinwari, Naveed. *Understanding Fata: Attitudes Towards Governance, Religion & Society in Pakistan's Federally Administered Tribal Areas Volume V.* Islamabad: Community Appraisal & Motivation Programme, 2012.

Singh, Ritika. 2013. "Drone Strikes Kill Innocent People. Why Is It so Hard to Know How Many?" *New Republic*. Accessed May 7, 2019. https://newrepublic.com/article/115353/civilian-casualties-drone-strikes-why-we-know-so-little

Slaughter, Anne-Marie. 2019. "Celebrating 20 Years of New America: New Ideas and New Voices to Renew America." *New America*. Accessed May 2, 2019. https://www.newamerica.org/new-america/reports/celebrating-20-years-new-america/

South Asia Terrorism Portal. 2019. "Fatalities in Terrorist Violence in Pakistan 2003–2019." *Institute for conflict management*, June. Accessed June 18, 2019. http://www.satp.org/satporgtp/countries/pakistan/database/casualties.htm

Taj, F. 2010. "Drone Attacks: Challenging Some Fabrications." *The Daily Times*, January. Accessed March 11, 2019. http://dailytimes.com.pk/opinion/01-Jan-10/drone-attacks-challenging-some-fabrications

Tariq, Sarmad. 2013. "Drone Strikes Kill Militants, Others Kill the Innocent." *The Express Tribune*, January. Accessed April 13, 2019. http://blogs.tribune.com.pk/story/15697/drone-strikes-kill-militants-others-kill-the-innocent/

TBIJ. 2018. "Strikes in Pakistan." *The Bureau of Investigative Journalism*, July. Accessed March 10, 2019. https://www.thebureauinvestigates.com/projects/drone-war/charts?show_casualties=1&show_injuries=1&show_strikes=1&location=pakistan&from=2004-1-1&to=now

Waraich, Omar. 2013. "Cricketer Turned Politician Imran Khan Leads Anti-Drone Protesters in Blockade of Key Nato Supply Line." *The Independent*, November. Accessed May 26, 2019. http://www.independent.co.uk/news/world/asia/cricketer-turned-politician-imran-khan-leads-anti-drone-protesters-in-blockade-of-key-nato-supply-8959637.html

Williams, Brian Glyn. "New Light on Cia "Double Tap" Drone Strikes on Taliban "First Responders" in Pakistan's Tribal Areas." *Perspectives on Terrorism* 7, no. 3 (2013): 79–83.

Yousaf, Farooq. 2013. "Drone Strikes: For Better or for Worse?" *The Express Tribune*, June. Accessed March 19, 2019. https://blogs.tribune.com.pk/story/17551/drone-strikes-for-better-or-for-worse/

Yousaf, Farooq. 2014. "The Drone Strike Narrative in Pakistan." The Diplomat, June. Accessed May 11, 2019. https://thediplomat.com/2014/06/the-drone-strike-narrative-in-pakistan/

Yousaf, Farooq. *CIA Drone Strikes in Pakistan: History, Perception and Future*. Islamabad: Center for Research and Security Studies, 2017. DecemberAccessed March 12, 2019. https://papers.ssrn.com/sol3/papers.cfm?abstract_id=3160433

Yousaf, Farooq, and Rahmanullah. "Drone Strikes in Fata: A Violation of Pakistan's Sovereignty." *Tigah: A Journal of Peace and Development* V (2014): 117–132.

Yousaf, Farooq, Haroon Rashid, and Imtiaz Gul. *Fata Tribes: Finally Out of Colonial Clutches? Past, Present and Future*. Islamabad, Pakistan: Center for Research and Security Studies, June 2018. Accessed April 10, 2019. https://www.researchgate.net/publication/326352642_FATA_Tribes_Finally_Out_of_Colonial_Clutches_Past_Present_and_Future

Zenko, Micah. 2016. "Do Not Believe the U.S. Government's Official Numbers on Drone Strike Civilian Casualties." *Foreign Policy*, July. Accessed June 19, 2019. http://foreignpolicy.com/2016/07/05/do-not-believe-the-u-s-governments-official-numbers-on-drone-strike-civilian-casualties/

What's in it for us? Armed drone strikes and the security of Somalia's Federal Government

Brendon J. Cannon

ABSTRACT

Strikes conducted by the US using armed unmanned aerial vehicles (UAVs) against the terrorist group al-Shabaab have received widespread attention documenting the number of strikes, the reactions of the local Somalis, and the disparity between AFRICOM and local reporting. Using interview data from individuals in Mogadishu, Nairobi and elsewhere in the region, this article provides a counterpoint to previous studies by measuring the Federal Government of Somalia's (FGS) reliance on the US drone campaign in Somalia in order to maintain its tenuous hold on power. It measures the extent to which the drone campaigns of external parties are able to insulate weak host governments from the threat of an internal foe. It finds that US drone strikes offer the occasional but only reliable check on al-Shabaab thereby allowing the FGS to continue functioning in Mogadishu. Yet strikes should not be construed as a sign of overt US support for the FGS. Rather, the US government, deeply suspicious of a highly corrupt and incapable FGS, continues to prosecute drone strikes against al-Shabaab because Washington considers the group poses a national security threat to the US.

Introduction

In September 2014, drones fired several Hellfire missiles and dropped various other munitions on a remote location in southern Somalia. The location targeted reportedly housed militants associated with the al-Shabaab terrorist group. The strike, confirmed by US Africa Command (AFRICOM),[1] openly acknowledged the use of drones in the attack and announced that the strike had successfully killed Mukhtar Abu Zubeyr, AKA Ahmed Abdi Godane, al-Shabaab's leader and reported mastermind of numerous lethal terrorist attacks that included the 2013 Westgate shopping mall attack in Nairobi, Kenya.

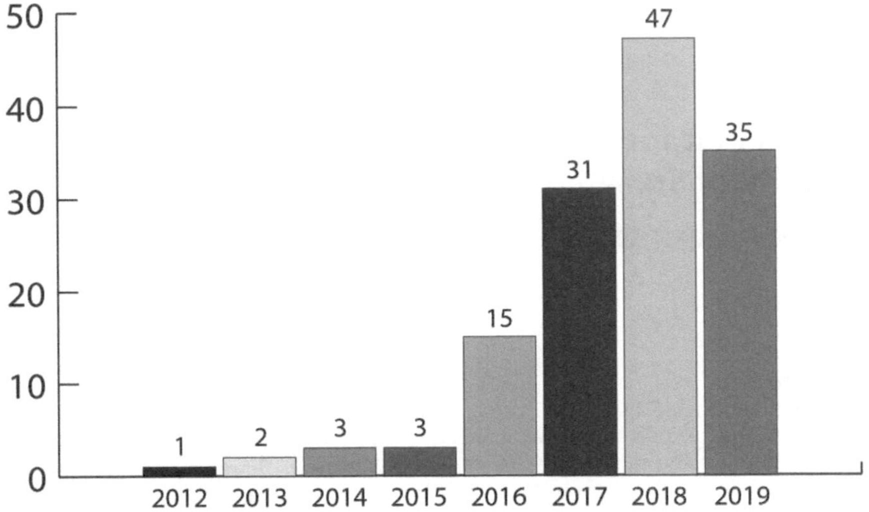

Figure 1. US drone strikes in Somalia, 2012–2019. Source of information: FDD's Long War Journal. https://www.longwarjournal.org/us-airstrikes-in-the-long-war.

Armed US drones have reportedly been used in Somalia against al-Shabaab since early 2007 when a predator drone was used to track an al-Qaeda (AQ) operative's convoy. Drones were first used as a separate attack medium in June 2011 and have steadily increased over time (see Figure 1). Between mid-2017 and mid-2019, after US President Donald Trump intensified the use of drones by designating parts of Somalia an 'area of active hostilities,' 110 strikes in Somalia were reported by AFRICOM.[2]

One reason drone strikes are used by states is to drastically limit the number of potential casualties resulting from combat because no physical fighting occurs during a strike. At the same time drones, as used by states as well as some non-state armed actors, constitute one of the most effective and lethal method of hampering the ability of the enemy to communicate, move and attack.[3] Drone strikes in Somalia against al-Shabaab appear to limit their control of territory and critical nodes of power projection for a time and may also hamper their ability to launch domestic, transborder or extra-regional attacks.

Given the importance, purported effectiveness and controversy surrounding US Government (USG) drone strikes in Somalia, this article explores the extent to which the Federal Government of Somalia (FGS) relies on kinetic strikes by US drones for survival, security, longevity and the resultant opportunity spaces they afford to Somalia's political elite. In essence, it begins to answer much broader questions about the extent to which the drone campaigns of external parties are able (or not) to insulate weak host governments from internal threats. In other words, the drone strikes may keep them from the very real possibility of outright collapse in the face of a sustained and tenacious internal foe. This article attempts to answer these pressing questions by first reviewing past and ongoing drone campaigns and literature related to them. Second, an analysis of US drone strikes, various missions and outcomes in Somalia against al-Shabaab is offered. Third, drone strikes are contextualized within the general state of security in Somalia, paying special attention to the congruence and divergence of USG and FGS interests. The fourth section examines how US drone strikes are operationalized in Somalia's unique political and security context. The final section demonstrates the USG will continue to launch drone strikes because of their effectiveness in downgrading what is deemed a US national security threat and, relatedly, that the security dividends accrued by the FGS on account of the strikes are of secondary interest to Washington.

Because this article focuses exclusively on the connection between US drone strikes and the survival of the FGS, it largely eschews references to the African Union Mission in Somalia (AMISOM),[4] its security operations and role, and the variety of other security actors, to include clan militias, operating in Somalia. Relatedly, it uses publicly available information about US strikes as published by AFRICOM and avoids referencing drone strikes in Somalia possibly conducted by the US Central Intelligence Agency (CIA), because they remain unsubstantiated.

Effectiveness, lethality and collateral damage

A drone strike is an attack carried out by an armed unmanned aerial vehicle (UAV), more popularly known as drones. UAV strikes generally include firing a missile or releasing a bomb at a human or other type of target. The drone may be equipped with such weapons as guided bombs, cluster bombs, incendiary devices, air-to-surface missiles, air-to-air missiles, anti-tank guided missiles or other types of precision-guided munitions.

Airstrikes from military-grade drones, reportedly used by a number of countries such as the US, China, Russia and Israel, are highly lethal. Drone-related literature has mushroomed since 9/11, precisely because the US along with a few other states have increasingly turned to drones to prosecute military action against a variety of armed groups spread across the globe.[5] Whether they are effective in countering insurgencies and armed groups in the long run is the subject of debate. Much of the literature on drones attempts to address this divide. It has done so alternatively by providing numbers of strikes, numbers of insurgents killed and target locations as well as demonstrating an increase or decrease in the lethality of armed attacks against locations and individuals targeted by the armed groups.[6] Some have found that targeted killing via drone strikes results in a decrease in violence because they hampers the capabilities of armed insurgent groups.[7] For example, Johnston and Sarbahi found that drone strikes correspond with changes in behavior by terrorist groups and, relatedly, a reduction in the incidence and lethality of terrorist attacks.[8] The opposite results, however, have been found by other scholars such as Abrahms and Mierau, Walsh and Smith, and Lehrke and Schomaker.[9] Other research has delved into collateral damage and what role this may play in either increasing or decreasing insurgent ranks.[10] Yet another genre has explored domestic factors within states prosecuting the drone strikes, arguing that they are effective precisely because they allow states with military drone capabilities to avoid expending significant amounts of blood and treasure in combat operations.[11] But the alternative viewpoint is also widely argued, with some scholars positing that drone strikes are only effective in the short term in countering insurgencies because no combat soldiers are present to capture and hold territory.[12] Legal aspects have also formed a large part of the literature, with scholars such as Strawser as well as Aloyo arguing that the assassination of powerful individuals by drone strikes is a moral imperative in order to avoid a wider conflict that would lead to more causalities.[13] The research of both McNeal and Etzioni supports this, demonstrating that drone strikes may lead to a reduction in civilian as well as military deaths.[14] Renic, on the other hand, finds the increased proclivity of some states to use drones 'is moving us ever closer to a mode of violence that strains the basic foundations upon which our moral justifications for killing in war rest.'[15]

In a geographic sense, the vast majority of the literature has focused on the US drone campaign against AQ and related Islamists as well as the Taliban along the Afghanistan/Pakistan border. Indeed, this campaign and its disputed results form the largest body of literature on military strikes by armed drones.[16] Less common, but nonetheless important, is the literature related to drone strikes in places far removed from the Hindu Kush. These include articles and analyses about US drone strikes and operations in Libya, Iraq, and Yemen.[17] Indeed, literature and analyses often quickly follow the onset of an armed drone campaign. For example, after the US and France confirmed they both were operating drones based in Niger in 2018, one scholar highlighted that the complexity of Niger's insurgent movements would necessarily hamper the effectiveness of armed drones.[18] Thus, is it all the more surprising that the subject of drone strikes in Somalia, where drone strikes in Somalia, which have been carried out for over a decade and with increasing frequency, is vastly under-researched.[19]

Much less effort has been spent by scholars attempting to catalogue and measure what effects drone strikes may have on host country governments.[20] That is, to what extent do drone strikes insulate governments from collapse in weak or failed states? In this case, collapse does not mean the inability of a government to govern effectively. Rather, it means the government –

regardless of capacity and/or intent to govern – avoids being removed from power by hostile actors such as a non-state armed group. Accordingly, this article provides a much-needed corrective to previous studies by using the case study of Somalia and measuring the reliance of the FGS on strikes conducted by US drones in order to maintain its tenuous hold on power.

Supporting the mission: U.S. drone strikes in Somalia

UAVs are operated in Somalia by the US Joint Special Operations Command (JSOC).[21] JSOC reportedly has its own fleet of armed Reaper drones, which are flown from various US bases in Djibouti, Kenya, Ethiopia and from within Somalia itself.[22] In addition, JSOC troops are reportedly deployed from time to time on the ground for surveillance, reconnaissance, and assault and capture operations.[23]

US drones strikes in Somalia target al-Shabaab, also known as *Harakat al-Shabaab al-Mujahidin*.[24] This terrorist group originated as the largely autonomous, militant wing of the Somali Council of Islamic Courts (CIC) that took over most of southern Somalia in the second half of 2006. After its military defeat, the CIC splintered into smaller factions of which al-Shabaab emerged as the most visible offshoot, intent on ostensibly establishing a theocracy across the Horn of Africa and East Africa. Its leaders at the time were assessed to be affiliated with AQ and are believed to have trained and fought in Afghanistan. The merger of these two terrorist groups was publicly announced in February 2012 by the amir of al-Shabaab and Ayman al-Zawahiri, leader of AQ. There are indications that al-Shabaab has recently split and that some fighters are now affiliated with the Islamic State (IS), also known as Daesh.

The group commits frequent terrorist attacks and not only maintains territorial control over parts of Somalia, but possesses territorial ambitions; i.e. it wishes to establish control over the entire state of Somalia and regions in adjacent states.[25] 'Al-Shabaab's primary objective is the establishment of an Islamist state in the Horn of Africa based on Sharia law and the elimination of secular and foreign influence, including through violent means … Al-Shabaab continues to conduct attacks in Somalia to achieve its aims. It primarily targets Somali Government interests … using mortars, rocket-propelled grenades, improvised explosive devices (IEDs) and firearms.'[26] Accordingly, al-Shabaab is both a Somalia-based and trans-border terrorist group.[27] That is, it commits crimes, attacks and atrocities within and without Somalia and has a proven track record of attacks – including against US interests and personnel – across the Eastern Africa region, most notably in Kenya.[28] Most importantly for the purposes of this article, it is considered by Washington to pose a serious and ongoing threat to the US homeland and US interests abroad.[29] The US targets al-Shabaab under the 2001 Authorization for the Use of Military Force (AUMF), a law that grants the US president the right to target and kill AQ and its affiliated group members globally. An August 2018 AFRICOM press release, for example, about a US drone strike in Somalia noted, 'US forces will continue to use all authorized and appropriate measures to protect U.S. citizens and to disable terrorist threats. This includes partnering with … Somali National Security Forces (SNSF) in combined counterterrorism operations and targeting terrorists, their training camps, and their safe havens throughout Somalia and the region.'[30]

US drone strikes in Somalia support a number of different missions. First, some strikes target al-Shabaab bases. Second, drones are used to target senior al-Shabaab officials, such as the aforementioned Godane in 2014. These have been referred to as VIP killing missions.[31] A third type of drone activity, which may include strikes, is tactical and provides support for troops on the ground. These could be done in support of US Special Forces operating in Somalia or in support of the military operations of AMISOM or those of troops ostensibly under the command of the FGS. Lastly, during routine surveillance missions, drones may spot al-Shabaab operatives or leaders and a strike is then initiated. These strikes are termed random targets of opportunity.[32]

The four types of drone strikes support a common goal: the downgrading and destruction of al-Shabaab operatives, leaders and materiel so they cannot launch attacks against the US homeland as well as US interests. The strikes also hamper the group's ability to attack targets in Somalia, particularly the FGS, and the Horn of Africa and East Africa regions. Examples of the lethality of drone strikes include a US operation involving drones and air strikes several miles northwest of Mogadishu in March 2016 that reportedly killed approximately 150 fighters; a large operation in November 2017 which killed more than 100 suspected Shabaab militants; and a strike in January 2019 in Jilib that killed between 52–73 militants.[33] In 2016, the US targeted an al-Shabaab leadership figure, Hassan Ali Dhoore, because of his involvement in the planning and execution of attacks against the Mogadishu airport, a hotel in the capital and the assassination of at least three American citizens. The drone strike, therefore, removed a leader responsible for attacking US interests and those of the FGS and demonstrated the two are not necessarily mutually exclusive. According to a Pentagon spokesperson, 'Removing Dhoore from the battlefield would be a significant blow to al-Shabab's operational planning and ability to conduct attacks.'[34] In three separate press releases from 2019 after drone strikes in Somalia, US commanders at AFRICOM struck a similar tone in explaining the rationale for US drone strikes: degrading al-Shabaab's ability to attack the US and kill Americans also degrades its ability to attack the FGS and kill Somalis. According to one statement, 'Precision airstrikes such as this one maintain pressure on al-Shabaab and disrupt its planning cycle and degrade its ability to mass forces and coordinate attacks against the Somali people.'[35] The next stated, 'This precision airstrike successfully targeted an encampment from which al-Shabaab could stage, coordinate, and execute attacks ... '[36] The third noted, 'In the case of this strike, we interrupted an Al Shabaab attempt to mass their forces. Precision strikes eliminated a potential threat to our partners and to the people of Somalia well before the terrorists ever got themselves organized.'[37]

Some have questioned the veracity of the statements issued by the Pentagon and AFRICOM, particularly in the face of reports of collateral damage and civilian casualties.[38] It is impossible, however, to question the veracity of al-Shabaab's stated aims and their related violent actions that have left thousands dead over the past 15 years. Their ultimate aim is state capture. There is little reason to doubt that al-Shabaab, if left unmolested by US drone strikes, would not attempt to fulfill its aim of removing the FGS and its foreign supporters. This is not hypothetical. The CIC – al-Shabaab's forerunner – forcibly took the capital and much of the country, ruling for a short time. While the CIC's rule brought a modicum of stability after years of chaos, this came at a steep price and, as the CIC became politicized and splintered, extortion, arbitrary justice, murders and violence became the order of the day. Al-Shabaab was and remains the most radical and militant offshoot of the CIC. Thus, while some have created the aura a 'golden age' of stability during the CIC's rule of Mogadishu, there is little supporting evidence for this and even less to believe this would be the outcome under the leadership of al-Shabaab.[39]

Decisions to increase drone strikes over the past decade indicate that successive US presidents and their advisors agree with many analysts, Somali officials and even al-Shabaab defectors that US drone strikes 'have become one of the most effective tools in confronting [al-Shabaab]. The air campaign has hindered al-Shabaab's ability to communicate, sown widespread mistrust among its members, and restricted its leaders' mobility.'[40] Drone strikes coupled with AMISOM operations have also reportedly resulted in al-Shabaab abandoning or losing territory, at least for a time.[41] As such, they are clearly an effective tactic used by the USG to partially degrade the capabilities of al-Shabaab. While drones can have a devastating effect on the group's ability to turn the tide in its favor they are, however, unable to remove al-Shabaab from the body politic of Somalia. After many years of drone strikes, al-Shabaab remains a serious threat across large parts of Somalia and as well and neighboring states.[42]

The state of Affairs in Somalia

Somalia is among the most emblematic of failed states. It has been without a functional government that can broadcast its power beyond certain urban areas for three decades. Indeed, the various governments of Somalia had no control of the capital until al-Shabaab's removal in 2011. Even today, the FGS does not maintain a monopoly on the use of force, legitimate or otherwise, in the capital. Instead, as discussed below, various political and clan factions jockey for control of critical infrastructure, neighborhoods, buildings, materiel and security personnel. In short, and despite rosy prognostications dating back over a decade that Somalia has 'turned a corner,' the collapse of the FGS is remains a distinct possibility given the multiplicity of internal threats facing it.[43]

While there are certainly disparities between the preceding Transitional Federal Governments (TFG)[44] and subsequent FGSs in terms of funding, capabilities and territory, all can be characterized as being either 'unable and/or unwilling to provide for the essential human needs of its citizens – in terms of security as well as adequate food, clean water, health care, and education.'[45] Rules that largely govern the international relations and domestic affairs of most sovereign states are largely absent in Somalia. Accordingly, the security situation – one that features a weak, corrupt government characterized by infighting and an armed, Islamist insurgency intent on establishing its control over Somalia – is regularly described as either 'fragile' or 'broken.'

Multiple factors influence and partially explain the state of affairs in Somalia despite years of funding, capacity building efforts, drone strikes and military operations.[46] Somalia's body politic, for example, is highly corrupt and the country remains at the top of the list of the world's most corrupt countries.[47] Corruption reportedly infects all levels of the public and private sectors. It is visible and expected. Finances are pilfered and squirreled away by those possessing political power rather than being used to build or expand state institutions.[48] Land belonging to government ministries, schools or other organizations has been reportedly seized by various politicians 'for security reasons.'[49] Paying taxes to a corrupt government is reportedly scrupulously avoided, though unpaid tax officials themselves are not about to expend the effort to find offenders.[50] Bribery across business and government sectors is the norm and corrupt practices such as nepotism are regularly associated with government procurement of contracts, which are often awarded to relatives, friends and associates of leading political figures or the external firm or organization offering the highest bribes.[51] In addition, Somalia's government institutions were either destroyed during the Somali Civil War, have been progressively starved of funding or repeatedly cannibalized by their leaders.[52]

To be clear, the FGS has both the mandate and the desire to assert its power from Mogadishu to the periphery. This is based on precedent. Somalia, unlike many other African states, actually has a history of strong governments and firm control emanating from the center to the periphery.[53] It has not always been a failed state and many Somali well remember the time when their passport was readily accepted internationally and travel over newly paved roads from Kismayo and Hargeisa to Mogadishu was not only possible, but a rite of passage from the periphery to the center for students and bureaucrats. These memories, however, are fading given the inability of the FGS to claw back control of even the capital territory, let alone the various federal states.[54] Sadly, Somalia has been a failed state (1991-present) for almost as long as it was not (1960–1990).

Despite the failure of the FGS to reform and rebuild the state, generous dollops of aid by external actors have largely continued. This can be chalked up to reasons of path dependency as well as the repeated humanitarian crises that wash over Somalia.[55] But any analysis of continuing assistance to Somalia must take into account what can only be understood as the willful ignorance of both state and non-state donors. This is not simply a problem of brazen corruption and a related inability or unwillingness to foster state institutions and governance on the part of the FGS. Rather, one must also take a hard look at the litany of 'capacity building' schemes that involve bribery, corruption, less-than-transparent contracts, graft and enrichment on the part of those donors and their related organizations.[56] In addition, decades of external funding for everything from food to security to government institutions has also seen external state and non-state actors embed themselves within the body politic of Somalia. According to one Somali scholar:

> Somalia continues to be bedeviled by a toxic cocktail of tribalized zones, self – seeking individualism, fissiparous identity politics, corruption as a way of life among the lumpen elite, and direct and dark financial interventions by foreign countries, particularly from the Middle East and some EU countries in search of compradors. This condition, now entering its third decade, gives the lie to the claim of the existence of an effective government in Somalia. On the contrary, the writ of Mogadishu is not uncontestably enforceable in the whole of the capital, let alone maintaining law and order across the width and breadth of Somalia.[57]

Despite being a ubiquitous feature and concept in international relations, there is little to no consensus as to how best to measure influence and its derivative concept power.[58] Yet in Somalia, there is ample evidence of the influence of external states such as the US, Turkey, Qatar, Italy and the UK. These states and others have become highly political actors as they have attempted to sway outcomes from presidential elections to oil sales and fisheries licenses to security sector reform (SSR) and infrastructure development.[59]

Given the state of affairs, the security situation in Somalia continues to deteriorate and the FGS shares much of the blame. It is neither a unitary actor nor does it exert control as a unified government over say the military or police, such as they exist.[60] Instead, FGS leaders are divided for reasons related to personal and family enrichment, clan and sub-clan, as well differences between diaspora Somali (often possessing dual citizenship) versus non-diaspora Somali (those who stayed in Somalia). In this volatile situation:

> Risk management strategies dominate local decision-making and can result in patterns of collusion, commoditization of security, and reliance on communal forms of protection and governance that work against efforts to build the formal security sector and state-based rule of law.[61]

In other words, this is a highly contested rather than an ungoverned space and Somalia's political actors accordingly attempt to leverage the firepower and manpower of the various security elements such as police or paramilitary units to further boost their own power and access to scarce resources.[62] What worries some Somalis, given the magnitude of threats coupled with the weak and ineffectual government, is not if the FGS will collapse, but when.

This point may best be illustrated by analyzing the list of FGS leaders in Table 1. The names of these individuals as well as some other perennial officeholders such as Omar Abdirashid Ali Sharmarke[63] are well-known to anyone who has studied the political situation in Somalia for the past two decades. That is, Farmajo, Beyle and others have all been part of or associated with the TFG and successive FGSs for years. They are part of a larger group of 'lumpen elite' who successfully harness Somalia's uniquely unstable political and security situation to remain in power and enrich themselves in the process. In turn, they frequently rotate bureaucrats and other appointees 'to regulate and control rent seeking, to prevent rivals from developing their own power base, and to demonstrate their power.'[64] A 2018 report that detailed government violence in Somalia is telling:

> Government security forces and allied militias, other persons wearing uniforms, regional security forces ... committed arbitrary or unlawful killings. Government and regional authorities executed persons without due process. Armed clashes and attacks killed civilians and aid workers ... Impunity remained the norm.[65]

Exacerbating this already combustible situation, most military and police units have been armed and trained over the course of the past decade by various external state actors such as Kenya, the United Arab Emirates (UAE) or the US; or international bodies such as the EU. These capacity building and security sector reform (SSR) measures meant to bolster security in Somalia have instead largely reified divided loyalties along clan or sub-clan lines as well as having added another layer of division in the form of foreign backing and the competing agendas of those state actors.[66] This situation engenders and encourages a profound insecurity in the capital that has ripple effects throughout the country. Thus, rather than expanding and ensuring security in Somalia, SSR and related funding are used to train security personnel who are either (mis)used by the FGS and/or operate independently to push various clan-based and resource-grabbing agendas thereby further undermining and undercutting the security situation.[67]

Into this maelstrom the omnipresent threat of al-Shabaab must be added (see Figure 2). The group has perpetrated attacks across the length and

Table 1. Current FGS leaders with their various allegiances, to include current FGS positions, former employers, clan allegiances and citizenship(s).

Name	Other name(s)	Position/Former position(s)	Clan	Sub-Clan	Sub-Sub-Clan	Citizenship(s)
Mohamed Abdullahi Mohamed	Farmajo	President of the Republic of Somalia	Darod	Marehan	Reer Dini	Somalia[a]
Hassan Ali Khaire	Hassan Khayre	Prime Minister/Regional Director of the Norwegian Refugee Council; Director of the British oil company Soma Oil	Hawiye	Murusad	Foor-Ulus	Norway and Somalia
Fahad Yasin Haji Dahir	Fahad Yasin	Deputy Commander of NISA/Chief of Staff of Villa Somalia; journalist for Al-Jazeera	Hawiye	Sheikhal	Reer Aw Hassan	Kenya and Somalia
Jamal Mohamed Hassan	Gamal Mohamed Hassan	Minister of Planning, Investment and Economic Development/ Ambassador of Somalia to Kenya	Darod	Warsangale	Dubeys	Canada and Somalia
Ahmed Isse Awad		Minister of Foreign Affairs and International Cooperation	Darod	Majertein	Mohamud Suleiman	Canada and Somalia
Abdirahman Duale Beyle	Abdirahman Duale Beihleh	Minister of Finance/Minister of Foreign Affairs	Dir	Gadubursi	Mohamed Ase	USA and Somalia

[a]Farmajo renounced his US citizenship in August 2019 despite the country's constitution allowing dual citizenship, Farmajo lived and worked in the US for decades before returning to Somalia in 2010 when he served as Prime Minister for a brief spell (Muia, 2019).

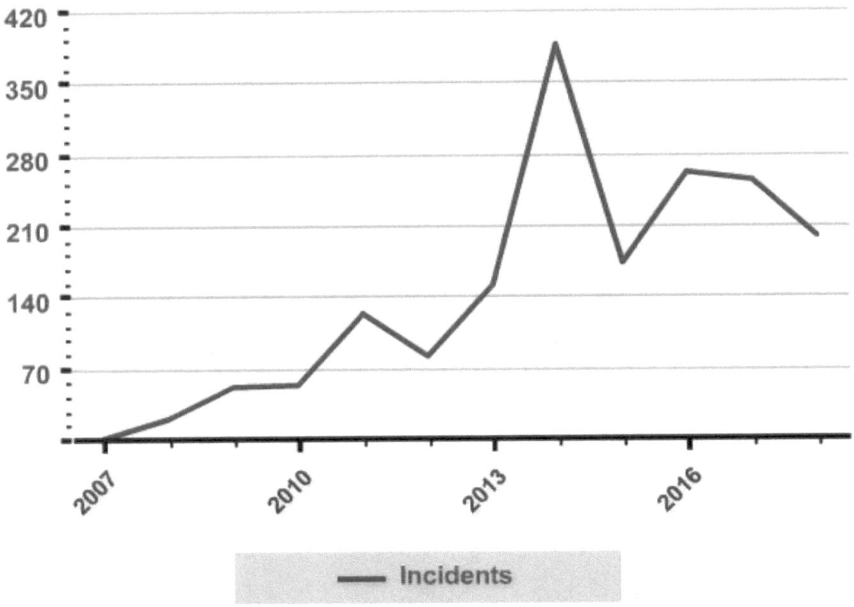

Figure 2. Number of al-Shabaab perpetrated terrorist attacks in Somalia, 2007–2018. The search results show attacks with all three Global Terrorism Database (GTD) criteria met and exclude ambiguous and unsuccessful attacks. Source: Global Terrorism Database (GTD), https://www.start.umd.edu/gtd/.

breadth of Somalia since 2007. In Mogadishu, on 15 June 2019, it launched two car bomb attacks in heavily secured government areas killing eleven. The previous month it detonated a car bomb, killing a Turkish citizen; carried out a suicide car bombing near president's residence, killing four civilians; and a carried out a car bombing at a security checkpoint that killed at least nine, including former foreign minister Hussein Elabe Fahiye. These attacks occurred in arguably the most secure city in the country: Mogadishu. The rest of the country, from Middle Shabelle to Puntland to Jubaland also saw numerous al-Shabaab attacks resulting in the deaths of civilians, soldiers and militants.[68] Ominously, security in Mogadishu has also been deteriorating during President Farmajo's term. After Farmajo's election in February 2017, there were over 120 armed attacks perpetrated in Mogadishu, including the massive bombing attack in central Mogadishu in October 2017 that resulted in 588 deaths.[69] 2019 began with an attack on the UN compound within the Mogadishu International Airport. The year ended with a car bomb targeting a tax center in Mogadishu that left 78 people killed and more than 100 wounded.

Al Shabaab not only remains a palpable and increasingly omnipresent threat, it seems resurgent, actively exploiting the security lapses and chaos engendered by FGS infighting and double dealing. With the exception of quarantined, US-trained troops, the security forces built up over the previous decades by various parties and states appear unable or unwilling to successfully counter the array of threats posed by al-Shabaab. If they are unable to protect themselves and the citizens of the capital, they are helpless in Somalia's hinterlands and cities further afield. The power accrued to al-Shabaab on account of the vacuum of power across much of the country also spills over into neighboring states. In early 2019, a headline-grabbing strike targeted an upscale hotel in Kenya's capital, Nairobi, 1700 km away from Mogadishu. In January 2020, three Americans were killed when al-Shabaab attacked a military base near Lamu, on Kenya's coast. These, perhaps more than anything else, continue to demonstrate the fecundity and longevity of al-Shabaab. They also underscored Washington's concerns that the group continues to pose a threat to the US and US interests and highlighted the fact that drone strikes pose one of the few credible deterrents keeping al-Shabaab from accruing enough power to eventually retake the capital.

Contextualizing US drone strikes in Somalia

It within this highly complex and fluid context that US drone strikes occur and increasingly target al-Shabaab across wide swathes of the country (see maps 1, 2 and 3). Yet, when references are made to the FGS and US drone strikes, scholars and journalists often feign shock when they point out the unilateral nature of US airstrikes and the fact that the FGS has little no say in where, when or how the strikes are prosecuted.[70]

In situations where there is an absence of 'host' state consent, the US has often justified the strikes by referencing self-defense. Some have argued this provides a lawful basis for the use of force within international law under Article 51 of the UN Charter.[71] However, this is not the case in Somalia and the current and previous FGSs, as well as their predecessors, the TFG, have reportedly offered the US their valid consent for drone strikes.[72] For example, AFRICOM regularly uses language in press releases such as 'All six airstrikes were conducted in close coordination with the Federal Government of Somalia … ' or 'In coordination with the Federal Government of Somalia, US forces conduct precision airstrikes such as this to foil terrorist plans, eliminate al-Shabaab terrorists and reduce the enemy's freedom of maneuver.'[73]

Despite the press releases from AFRICOM, the level of coordination that occurs between the USG entities responsible for planning and prosecuting drones strikes and their FGS counterparts remains highly questionable. Indeed, sources note that there exists little to no coordination between the USG and FGS about drone strikes.[74] There are a number of reasons why this is the case.

First, given extremely robust military and intelligence capabilities on a global level, the US maintains its own extensive set of assets, both on the ground and in the air, that assist it in prosecuting drone strikes in Somalia.[75] Indeed, the US

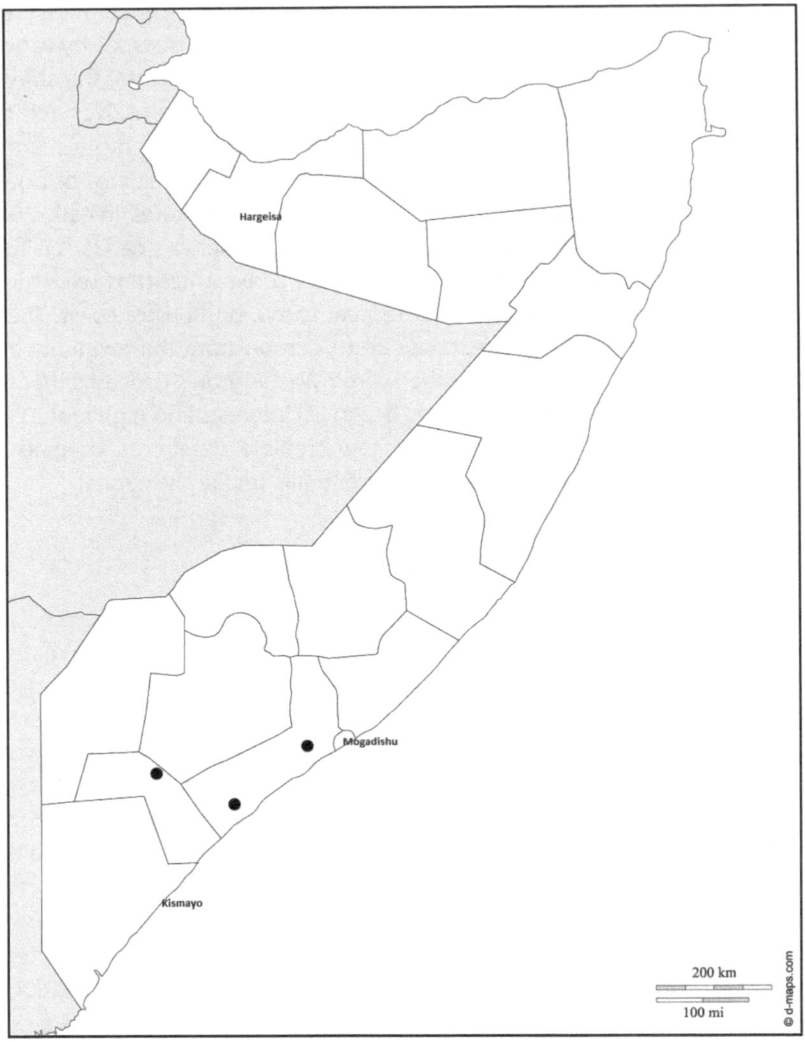

Map 1. Locations of US drone strikes in Somalia 2014. Source of information: FDD's Long War Journal. https://www.longwarjournal.org/us-airstrikes-in-the-long-war. Original map source: https://d-maps.com/carte.php?num_car=4044&lang=en.

has reportedly fostered, funded and trained various Somali state agencies that operate autonomously from the FGS.[76] Second, while valid and actionable intelligence on al-Shabaab movements and personnel from the FGS would likely be welcome, there are reportedly serious 'chain of acquisition' problems with intelligence provided by FGS-affiliated entities such as the National Intelligence Security Agency (NISA). In some cases, valid intelligence may be available, but payment or bribery is required to procure further details.[77] Third,

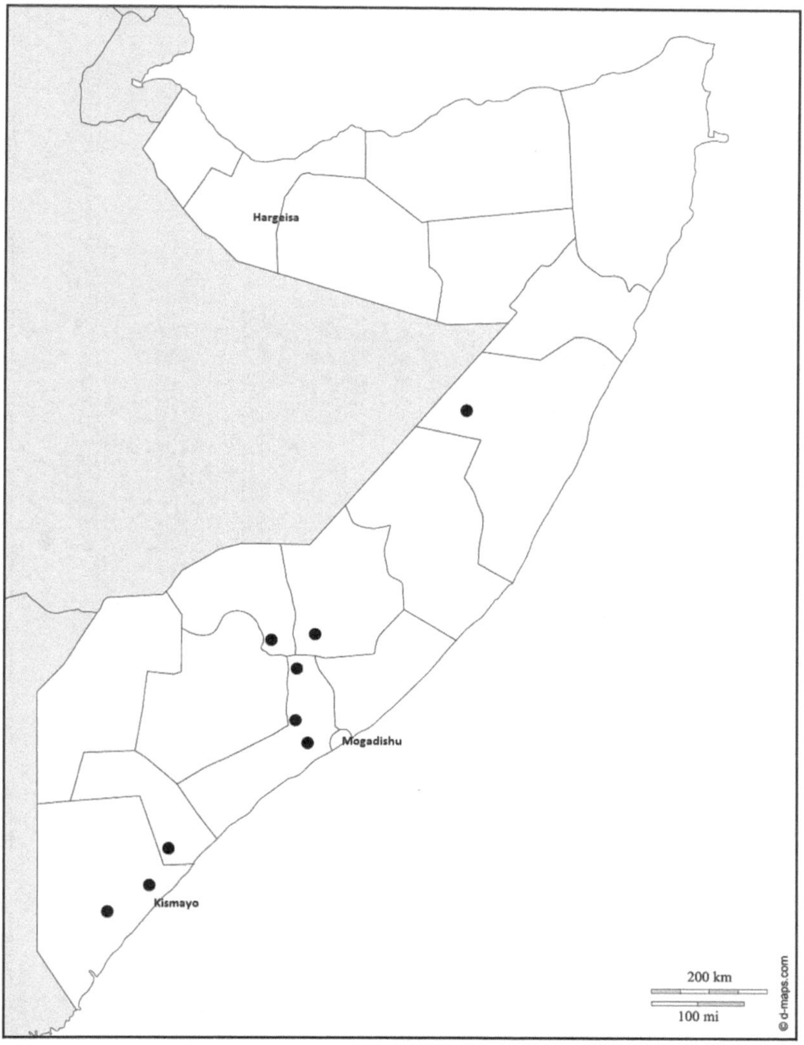

Map 2. Locations of US drone strikes in Somalia 2016. Source of information: FDD's Long War Journal. https://www.longwarjournal.org/us-airstrikes-in-the-long-war. Original map source: https://d-maps.com/carte.php?num_car=4044&lang=en.

many of the various intelligence and armed units associated (at least in name) with the FGS have repeatedly been infiltrated by al-Shabaab, thus further delegitimizing the FGS and its intelligence in the eyes of the Americans and other potential intelligence and military operations partners.[78]

Given the above, the USG have become deeply mistrustful of the FGS. Furthermore, it does not necessarily need the FGS for intelligence given

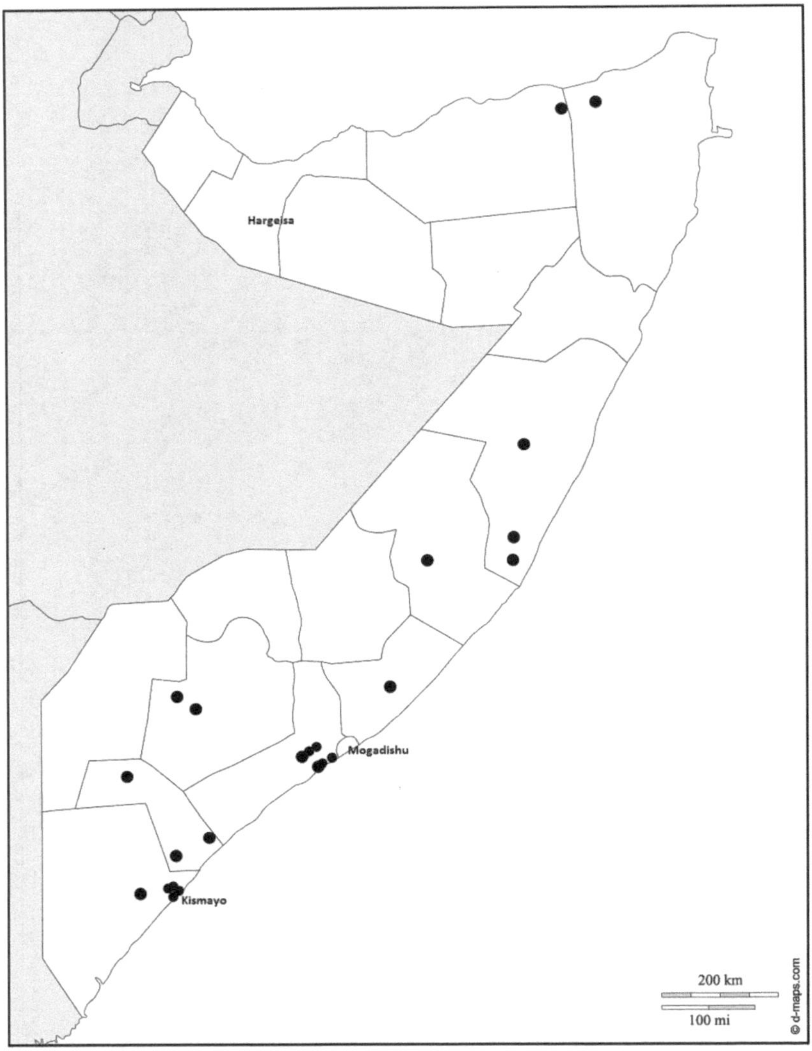

Map 3. Locations of US drone strikes in Somalia 2018. Source of information: FDD's Long War Journal. https://www.longwarjournal.org/us-airstrikes-in-the-long-war. Original map source: https://d-maps.com/carte.php?num_car=4044&lang=en.

robust and unilateral USG intelligence sources and methods. As such, there is little impetus for the US to coordinate or share information with the FGS about drone strikes. The AFRICOM press releases highlighting coordination with the FGS are therefore, more often than not, likely hyperbole. Indeed, the FGS may only learn of US drone strikes within Somalia after the fact via AFRICOM's announcement.[79]

US national security; FGS survival

US drone strikes in Somalia not only continue, but are on the rise – even though the USG long ago lost trust in the FGS. Strikes should be placed contextually within the situation of ongoing, extreme violence in both the capital and throughout large parts of the country as well as the corruption, impunity and impotence of the FGS in the face of such carnage. As the security situation deteriorates and al-Shabaab's influence and power waxes, the US continues its efforts to contain and degrade an AQ-affiliated terrorist group that Washington sees as posing a real and present national security threat. US drone strikes are therefore conducted as a way of destroying any credible threat posed by al-Shabaab to the US homeland, first and foremost, as well as US interests in the wider region.

Any security dividends accrued by the FGS are, therefore, secondary in nature. US drone strikes – as outlined above – offer the occasional but only reliable check on al-Shabaab thereby insulating the weak Somali government from the possibility of collapse in the face of a sustained internal threat. Strikes by armed UAVs are effective in countering the terrorist group's lethality because they are, in and of themselves, highly lethal. Simply put, drone strikes result in the deaths of terrorists, the destruction of their equipment and bases and the downgrading of their capabilities. This prevents them from successfully establishing their writ of power across the country by removing the FGS. That the strikes also result in civilian casualties, at times, is immaterial when answering the question posed here: do drone strikes of an external party effectively insulate weak host governments from internal threats? The answer to that question, based on the evidence compiled, is yes. There does appear to be a causal connection between drone strikes and the survival of the feeble FGS. The bitter irony is that FGS leaders, insulated from outright ruin, continue to maintain their positions of privilege and relative power at the expense of the vast majority of their citizens' safety, security and future while the threat of al-Shabaab, although constrained by drone strikes, stubbornly remains.

Notes

1. The United States Africa Command (USAFRICOM, U.S. AFRICOM, and AFRICOM), headquartered at Kelley Barracks, Stuttgart, Germany, is one of the eleven unified combatant commands of the United States Armed Forces.
2. An increase in the use of drones and other lethal strikes has occurred during the Trump administration, not only in Somalia but also in Yemen and Pakistan. Drone strikes numbered 238 between 2017 and the end of 2018, for example. This is compared with 186 strikes at the height of the Obama administration's drone use in 2009 and 2010. Pratt, *US Killing*. In addition, President Trump reportedly weakened the rules governing drone strikes outside declared war

zones, to include throwing out the requirement that a target pose an 'imminent threat.' Savage, *Will Congress Ever.*

3. Rossiter, 'Drone usage by militant,' 116–119.
4. Troops from various African states including Burundi, Uganda, Kenya and Ethiopia form AMISOM, a UN-sanctioned, African Union (AU)-operated force that has borne the brunt of fighting, capturing and holding territory since its inception in 2007. The actions of AMISOM as well as US drone strikes and other related security operations are largely responsible for limiting al-Shabaab's power and territorial acquisitions.
5. Milan and Tabrizi, 'The Need for UCAVs.'
6. Johnston and Sarbahi, 'The impact of US'; Price, 'Targeting Top Terrorists'; Reeder and Smith, 'US Strikes in Somalia'; and Jordan, 'When Heads Roll.'
7. Wilner, 'Targeted killings in Afghanistan'; and Dear, 'Beheading the Hydra'.
8. Johnston and Sarbahi, 'The impact of US.'
9. Abrahms and Mierau, 'Leadership Matters'; Walsh and Smith, 'Do Drone Strikes Degrade'; Lehrke and Schomaker, 'Kill, Capture, or Defend?'.
10. Shah, 'Do U.S. Drone Strikes Cause Blowback?'; Ahmed, Jenkins, and Iftikhar,"Perception of Foreign Drone Strikes by Citizens"; Jordan, 'Attacking the Leader'; and Mahmood, 'The Efficacy of US Drone Strikes in Pakistan.'
11. Rinehart, *Drones and Targeted*; Macdonald and Schneider, 'Presidential risk.'
12. Cockburn, *Kill Chain*; Carvin, 'Getting Drones Wrong'; and Cronin, 'Why Drones Fail.'
13. Strawser, 'Moral Predators'; and Aloyo, 'Just Assassinations.'
14. McNeal, 'Are Targeting Killings Unlawful?'; and Etizioni, 'The Great Drone Debate.'
15. Renic, 'Justified Killing,' 425.
16. Johnston and Sarbahi, 'The impact of US'; Aslam, "'A critical evaluation'; Franke, 'Drones, drone strikes'; and Mazhar, 'Do remotely piloted.'
17. Downes, 'Targeted Killings'; Eastin and Glade, 'Beheading the Hydra'; Pashakhanlou, 'Air Power'; and Barry 'Harsh Lessons.'
18. Guido, 'The American Way.'
19. One of the few exceptions beyond media stories or scholarly articles with passing references to Somalia is Reeder and Smith, 'US Strikes in Somalia.'
20. Arguments about the erosion (or not) of state sovereignty figure more prominently in the literature. But this research often presupposes a unitary and relatively robust state such as Pakistan, not a failed state like Somalia. Ali and Khan, 'Drone Strikes.'
21. Recent reports also claim drones and drone strikes in Somalia are operated by the US Central Intelligence Agency (CIA). However, given the nature of these claims and the opacity that generally characterizes CIA operations they do not figure in this article.
22. Hansen, *Somalia drone strikes*; and Byrne 'Consent and the use of force.'
23. There are currently 500 US troops reportedly deployed to Somalia. Morgan, *U.S. Military Builds Up*; and Purkiss and Fielding-Smith, *Somalia: Reported US Actions.*
24. In 2017, President Trump authorized the US military to conduct offensive strikes in Somalia against al-Shabaab fighters as well those affiliated with the Islamic State (IS), also known as the Islamic State of Iraq and Syria (ISIS).
25. Reeder and Smith, 591.
26. Australian National Security, 'Al-Shabaab.'
27. Cannon and Iyekekpolo, 'Explaining Transborder Terrorist Attacks.'

28. Cannon and Pkalya, 'Why al-Shabaab Attacks.'
29. For example, three Americans were killed when al-Shabaab fighters stormed a Kenyan air base used by the US military in Lamu county in early January 2020.
30. U.S. Africa Command Public Affairs, *U.S. Conducts Airstrike in Support.*
31. Hansen.
32. Ibid.
33. Bergen, *Drone Strikes: Somalia.*
34. BBC, 'Somalia Al-Shabaab.'
35. Purkiss and Fielding-Smith.
36. Ibid.
37. Ibid.
38. These seem to be couched in ethical or moral terms rather than based on an evaluation of the results of the strike in hampering the capabilities of al-Shabaab. In other words, the reports and related arguments maintain a focus on the lethal nature of drone attacks resulting in civilian casualties and collateral damage and generally ignore the lethality of al-Shabaab and what the group would do if left unhampered. See Amnesty International, 'The Hidden War in Somalia.' See also Sperber, *Inside the Secretive.*
39. For an excellent article documenting the stability, abuse and so-called 'golden age' of CIC rule in Mogadishu, see Barnes and Hassan, 'The Rise and Fall of Mogadishu's Islamic Courts.'
40. Goldbaum, *A Trumpian War.*
41. Ibid.; and Dörrie, 'Why Territorial.'
42. In January 2019, al-Shabaab attacked the upscale Dusit D2 hotel complex in the heart of Nairobi, Kenya's capital city, killing at least 21 people and injuring at least 28.
43. The adage 'Somalia has turned a corner' dates back well over a decade, but is supported by little evidence. Kachroo and Simmons, 'After Decades'; Kenyan, 'How Somalia's Government'; The Economist, 'Pushing It Across'; and Xinhua, 'AMISOM commits.' For a soberer analysis detailing the weakness and potentially imminent collapse of the FGS as compared with the continued power of al-Shabaab, see Ingiriis, 'Building Peace.'
44. The Transitional Federal Government (TFG) was the internationally recognized government of the Republic of Somalia from 2004 until 2012, when its tenure officially ended and the FGS was inaugurated.
45. Jamal, *Identifying Causes.*
46. Funding from external states and organizations has been and continues to be the lifeblood of the FGS. The US, for example, has spent a combined total of $2.9 billion since 2012. In 2016 alone, the US Department of Defense (DoD) spent over $67million under Section 1206, which allows for the training and equipping of foreign military forces to perform counterterrorism operations and to participate in or to support military and stability operations in which US Armed Forces are participating. Security Assistance Monitor, *Somalia.*
47. Transparency International, *2018 Corruptions Perceptions.*
48. Fartaag, Interview by Author.
49. Marqaati, *Somalia State of.*
50. Ibid.
51. Fartaag, Interview by Author; and Marqaati.
52. Fartaag, *Their Own Worst*; and Fartaag, *Breaking Point.*
53. Herbst, *States and power*; and Samatar, *Africa's First Democrats.*

54. The FGS exercises little to no control over vast swathes of Somalia. For example, the breakaway Republic of Somaliland – de-facto independent but internationally unrecognized since 1991 – signed a deal to develop the port of Berbera on the Gulf of Yemen with Ethiopia and the UAE's DP World in early 2016. Many in the FGS were incensed and subsequently passed laws that banned DP World from operating in Somalia. These made no difference and the port expansion continues as planned. See Cannon and Rossiter, 'Ethiopia, Berbera port.'
55. For examples of humanitarian aid path dependency, see Cannon and Fujibayashi, 'Security, structural factors.'
56. Much of the aid money from external states ends up being returned to contractors and subcontractors from those same states. They feel little to no pressure to complete projects or ensure their success because that would mean an end to the contracts. Other abuses of aid have been documented such as the UN World Food Programme (WFP)'s work with Somali cartels. Together they divided the lucrative profit derived from selling UN food aid. Jones, 'Half of all food'; and Mogadishu-based businessperson, Interview by Author.
57. Samatar, *Somaliland's Stakes.*
58. Keohane and Nye, 'Power and Interdependence'; Waltz, *Theory of International*; and Barnett and Duvall, *Power in Global.*
59. Cannon, 'Foreign State Influence'; and Hills, 'Security sector'; and Cannon and Donelli, 'Asymmetric Alliances.'
60. Reno, 'The politics of security,' 505.
61. Menkhaus, 'Managing risk.'
62. Mogadishu-based businessperson, Interview by Author. For an excellent analysis documenting the weakness and very real possibility of collapse of the FGS juxtaposed with the continued strength of al-Shabaab, see Ingiriis, 'Building peace.'
63. Prime Minister of Somalia from August 2009 until September 2010 and again from December 2014 until March 2017.
64. Balthasar, *State-making*, 131.
65. U.S. Department of State, 'Country Reports.' 2.
66. For example, after a diplomatic spat resulting from the Gulf Cooperation Council (GCC) crisis, at least 600 weapons were stolen from a former UAE-run training center Mogadishu. The training facility and arsenal were looted and the weapons were reportedly sold throughout the city. Sheikh and Omar, *Exclusive: Weapons stolen*.
67. Rossiter and Cannon, 'Re-examining the "Base",'181.
68. Crisiswatch, *Tracking Conflict.*
69. START, *Global Terrorism.*
70. Byrne; and Sperber, *Inside the Secretive.*
71. International Bar Association, 'The Legality of,' 13.
72. Byrne 'Consent and the Use of Force,' 120.
73. See note 35 above.
74. Mogadishu-based official, Interview by Author; and Fartaag, Interview by Author.
75. Mogadishu-based official, Interview by Author.
76. For example, an elite commando force known as 'Danab', or the lighting force, is reportedly one such unit that provides intelligence. Originally conceived as a battalion-sized element, AFRICOM reported that the effort has been expanded

to build a 3,000-troop brigade deployed across six Somali army sectors. Vandiver, *US helps Somali*; and Reno, 'The politics of security,' 508.

77. Somali intelligence officials reportedly warned Kenya in November 2018 about a plot to attack an unknown target involving five men – the same number who attacked the Dusit D2 hotel in Nairobi in January 2019. However, Kenya reportedly decided not to pay for more information because of the vague nature of the information as well as the sheer number of threats of attack by al-Shabaab routinely reported to Kenyan intelligence. Yet without payment, according to one Somali official, it was impossible to solicit further details from sources within al-Shabaab. Akwiri and Houreld, 'Kenya bomber's journey.'
78. For example, in 2013, attackers were able to infiltrate and attack both the heavily guarded and fortified prime minister's residence as well as Villa Somalia, the presidential palace. The wearing of badges and security service uniforms by al-Shabaab attackers is also indicative of major security lapses and infiltration. After their role came to light after a string of hotel bombings in Mogadishu, a scathing UN monitoring group report highlighted al-Shabaab infiltration of NISA, built and funded in part by the CIA, as well as the police force. Reno, 'The Dilemmas of Security,' 62; and United Nations, 64.
79. See note 62 above.

Acknowledgments

I would like to thank my good friend and colleague, Miguel Alejandro Híjar-Chiapa, for taking the time and effort to create a number of the tables and maps seen in this article. The author would like to thank Abdirazak Fartaag, former Head of Public Finance Management Unit, Office of the Prime Minister (Somalia) and founder of Fartaag Research and Consulting for his information, comments and suggestions. This research project would not have come to fruition without his assistance. The author would also like to thank his many friends in Hargeisa, Mogadishu and Nairobi who chose to remain anonymous but have helped him greatly with this and other endeavors.

Disclosure Statement

The author has no conflicts of interest to report.

ORCID

Brendon J. Cannon http://orcid.org/0000-0003-1731-9547

Bibliography

Abrahms, Max, and Jochen Mierau. "Leadership Matters: The Effects of Targeted Killings on Militant Group Tactics." *Terrorism and Political Violence* 29, no. 5 (2017): 830–851. doi:10.1080/09546553.2015.1069671.

Ahmed, Liban. 2016. *How to Fight Corruption in Somalia*. December 7. https://wardheernews.com/how-to-fight-corruption-in-somalia/.

Ahmed, Zahid Shahab, Bert Jenkins, and Waseem Iftikhar. "Perception of Foreign Drone Strikes by Citizens: The Context of US Drone Strikes in Pakistan." *South Asian Survey* 24, no. 2 (2017): 135–157. doi:10.1177/0971523118822442.

Ajala, Olayinka. "US Drone Base in Agadez: A Security Threat to Niger?" *The RUSI Journal* 163, no. 5 (2018): 20–27. doi:10.1080/03071847.2018.1552452.

Akwiri, Joseph, and Katharine Houreld. 2019. *Kenya Bomber's Journey Offers Cautionary Tale of Intelligence Failures*. January 31. https://www.reuters.com/article/us-kenya-security-bomber-insight/kenya-bombers-journey-offers-cautionary-tale-of-intelligence-failures-idUSKCN1PP1AP.

Akwiri, Joseph, and Katharine Houreld. 2019. *Kenya Bomber's Journey Offers Cautionary Tale of Intelligence Failures*. January 31. https://www.reuters.com/article/us-kenya-security-bomber-insight/kenya-bombers-journey-offers-cautionary-tale-of-intelligence-failures-idUSKCN1PP1AP.

Ali, Shahid, and Amjad Abbas Khan. "Drone Strikes, Logic of Self-defense and Violation of State Sovereignty." *Pakistan Journal of History and Culture* 38, no. 2 (2017): 187–206.

Aloyo, Eamon. "Just Assassinations." *International Theory* 5, no. 3 (2013): 347–381. doi:10.1017/S1752971913000237.

Amnesty International. *The Hidden War in Somalia*. London: Amnesty International, 2019.

Aslam, Muhammad W. "A Critical Evaluation of American Drone Strikes in Pakistan: Legality, Legitimacy and Prudence." *Critical Studies on Terrorism* 4, no. 3 (2011): 313–329. doi:10.1080/17539153.2011.623397.

Australian National Security. n.d.. *Al-Shabaab*. https://www.nationalsecurity.gov.au/Listedterroristorganisations/Pages/Al-Shabaab.aspx.

Balthasar, Dominik. *State-making in Somalia and Somaliland: Understanding War, Nationalism and State Trajectories as Processes of Institutional and Socio-cognitive Standardization*. London: Doctoral dissertation, The London School of Economics and Political Science (LSE), 2012.

Barnes, Cedric, and Harun Hassan. "The Rise and Fall of Mogadishu's Islamic Courts." *Journal of Eastern African Studies* 1, no. 2 (2007): 151–160. doi:10.1080/17531050701452382.

Barnett, Michael, and Raymond Duvall. *Power in Global Governance*. Vol. 98. Cambridge: Cambridge University Press, 2004.

Barry, Ben. *Harsh Lessons: Iraq, Afghanistan and the Changing Character of War*. London: Routledge, 2018.

BBC. 2016. *Somalia al-Shabab: US Drone Strike Targets Militant Chief.* April 2. https://www.bbc.com/news/world-africa-35949675.

Bergen, Peter, David Sterman, Melissa Salyk-Virk, Alyssa Sims, and Albert Ford. 2019. *Drone Strikes: Somalia.* https://www.newamerica.org/in-depth/americas-counterterrorism-wars/somalia/.

Byrne, Max. "Consent and the Use of Force: An Examination of 'Intervention by Invitation'as a Basis for US Drone Strikes in Pakistan, Somalia and Yemen." *Journal on the Use of Force and International Law* 3, no. (1) (2016): 97–125. doi:10.1080/20531702.2015.1135658.

Cannon, Brendon J. "Foreign State Influence and Somalia's 2017 Presidential Election: An Analysis." *Bildhaan: An International Journal of Somali Studies* 18 (2019): 20–50.

Cannon, Brendon J, and Federico Donelli. "Asymmetric Alliances and High Polarity: Evaluating Regional Security Complexes in the Middle East and Horn of Africa." *Third World Quarterly* (2019): 505–524. doi:10.1080/01436597.2019.1693255.

Cannon, Brendon J, and Hirotaka Fujibayashi. "Security, Structural Factors and Sovereignty: Analysing Reactions to Kenya's Decision to Close the Dadaab Refugee Camp Complex." *African Security Review* 27, no. 1 (2018): 20–41. doi:10.1080/10246029.2017.1408475.

Cannon, Brendon J., and Wisdom Iyekekpolo. "Explaining Transborder Terrorist Attacks: The Cases of Boko Haram and Al-Shabaab." *African Security* 11, no. 4 (2018): 370–396. doi:10.1080/19392206.2018.1560970.

Cannon, Brendon J, and Dominic R. Pkalya. "Why al-Shabaab Attacks Kenya: Questioning the Narrative Paradigm." *Terrorism and Political Violence* 31, no. 4 (2019): 836–852. doi:10.1080/09546553.2017.1290607.

Cannon, Brendon J, and Ash Rossiter. "Ethiopia, Berbera Port and the Shifting Balance of Power in the Horn of Africa." *Rising Powers Quarterly* 2, no. 4 (2017): 7–29.

Carvin, Stephanie. "Getting Drones Wrong." *The International Journal of Human Rights* 19, no. 2 (2015): 127–141. doi:10.1080/13642987.2014.991212.

Cockburn, Andrew. *Kill Chain: The Rise of the High-Tech Assassins.* New York: Henry Holt and Company, 2015.

Crisiswatch. 2019. *Tracking Conflict Worldwide.* June. https://www.crisisgroup.org/crisiswatch/print?page=3&location%5B0%5D=9&t=CrisisWatch+Database+Filter.

Cronin, Audrey K. "Why Drones Fail: When Tactics Drive Strategy." *Foreign Affairs* 92, no. 4 (2013): 44–54.

Dear, Keith Patrick. "Beheading the Hydra? Does Killing Terrorist or Insurgent Leaders Work?" *Defence Studies* 13, no. 3 (2013): 293–337. doi:10.1080/14702436.2013.845383.

Dörrie, Peter. 2016. "Why Territorial Losses Don't Weaken Somalia's Al-Shabab." *World Politics Review (WPR).* July 27. https://www.worldpoliticsreview.com/articles/19486/why-territorial-losses-don-t-weaken-somalia-s-al-shabab.

Downes, Chris. "Targeted Killings' in an Age of Terror: The Legality of the Yemen Strike." *Journal of Conflict and Security Law* 9, no. 2 (2004): 277–294. doi:10.1093/jcsl/9.2.277.

Eastin, Joshua, and Emily Kalah Glade. "Beheading the Hydra: Counterinsurgent Violence and Insurgent Attacks in Iraq." *Terrorism and Political Violence* 30, no. 3 (2018): 384–407. doi:10.1080/09546553.2016.1167688.

The Economist. 2014. "Pushing It across the Borders." *The Economist.* February 10. https://www.economist.com/middle-east-and-africa/2014/02/10/pushing-it-across-the-borders.

Etzioni, Amitai. "The Great Drone Debate." *Military Review* 93, no. 2 (2013): 2–13.

EU International Cooperation and Development. 2019. *"The European Union Steps up Its Support to Somalia with €200 Million at the London Conference."* https://ec.europa.eu/europeaid/news-and-events/european-union-steps-its-support-somalia-eu200-million-london-conference_en.

Fartaag, Abdirazak. *Their Own Worst Enemy: How Successive Governments Plundered Somalia's Public Resources and Why the World Looked On*. Nairobi: Fartaag Consulting, 2014. http://fartaagconsulting.com/publications.html.

Fartaag, Abdirazak. *Breaking Point in Somalia: How State Failure Was Financed and by Whom*. Nairobi: Fartaag Consulting, 2016. http://fartaagconsulting.com/publica tions.html.

Fartaag, Abdirazak, interview by Author. 2019 (January 6).

Felter, Claire, Jonathan Master, and Mohammed Aly Sergie. 2019. *Al-Shabab*. January 31. https://www.cfr.org/backgrounder/al-shabab.

Franke, Ulrike Esther. "Drones, Drone Strikes, and US Policy: The Politics of Unmanned Aerial Vehicles." *Parameters* 44, no. 1 (2014): 121–130.

Goldbaum, Christina. 2018. *A Trumpian War on Terror that Just Keeps Getting Bigger*. September 11. https://www.theatlantic.com/international/archive/2018/09/drone-somalia-al-shabaab-al-qaeda-terrorist-africa-trump/569680/.

Guido, L. T. C. Joseph. "The American Way of War in Africa: The Case of Niger." *Small Wars & Insurgencies* 30, no. (1) (2019): 176–199. doi:10.1080/09592318.2018.1554337.

Hansen, Stig Jarle. 2019. *Somalia Drone Strikes are a Potent Weapon, but Not the Game Changer*. February 20. http://theconversation.com/somalia-drone-strikes-are-a-potent-weapon-but-not-the-game-changer-111845.

Herbst, Jeffrey. *States and Power in Africa: Comparative Lessons in Authority and Control*. Princeton, NJ: Princeton University Press, 2014.

Hills, Alice. "Security Sector or Security Arena? the Evidence from Somalia." *International Peacekeeping* 21, no. 2 (2014): 165–180. doi:10.1080/13533312.2014.910400.

Hirsch, John L. 2011. "The Black Hawk Down Effect." *Foreign Policy*. https://foreign policy.com/2011/08/12/the-black-hawk-down-effect/.

Houreld, Katharine, and Abdi Sheikh. 2017. *Deadly Somalia Blast Reveals Flaws in Intelligence Efforts*. October 20. https://www.reuters.com/article/us-somalia-attacks/deadly-somalia-blast-reveals-flaws-in-intelligence-efforts-idUSKBN1CP20W.

Ingiriis, Mohamed Haji. "Building peace from the margins in Somalia: The case for political settlement with Al-Shabaab." *Contemporary Security Policy* 39, no. 4 (2018): 512–536.

International Bar Association. 2017. "The Legality of Armed Drones under International Law." Background paper by the International Bar Association's Human Rights Institute, adopted in May 2017, International Bar Association's Human Rights Institute, 1–56. https://www.ibanet.org/Human_Rights_Institute/council-resolutions.aspx.

Jamal, Ahmed Rashid. 2013. *Identifying Causes of State Failure: The Case of Somalia*. August 13. www.atlantic-community.org/-/identifying-causesof-state-failure-the-case-of-somalia.

Johnston, Patrick B, and Anoop K Sarbahi. "The Impact of US Drone Strikes on Terrorism in Pakistan." *International Studies Quarterly* 60, no. 2 (2016): 203–219. doi:10.1093/isq/sqv004.

Jones, Sam. 2010. "Half of All Food Sent to Somalia Is Stolen, Says UN Report." *The Guardian*, March 10. https://www.theguardian.com/world/2010/mar/10/aid-soma lia-stolen-un-report.

Jordan, Jenna. "When Heads Roll: Assessing the Effectiveness of Leadership Decapitation." *Security Studies* 18, no. 4 (2009): 719–755. doi:10.1080/09636410903369068.

Jordan, Jenna. "Attacking the Leader, Missing the Mark: Why Terrorist Groups Survive Decapitation Strikes." *Security Studies* 38, no. 4 (2014): 7–38.

Jowell, Marco. "Jowell, M. (2018). The Unintended Consequences of Foreign Military Assistance in Africa: An Analysis of Peacekeeping Training in Kenya." *Journal of Eastern African Studies* 12, no. 1 (2018): 102–119. doi:10.1080/17531055.2017.1418187.

Kachroo, Rohit, and Keir Simmons. 2013. "After Decades as 'World's Most Dangerous' Place, Has Somalia Turned the Corner?" *NBC News*. May 11. https://wardheernews.com/after-decades-as-worlds-most-dangerous-place-has-somalia-turned-the-corner/.

Keohane, Robert O., and Joseph S. Nye. "Power and Interdependence." *Survival* 15, no. 4 (1973): 158–165. doi:10.1080/00396337308441409.

Keynan, Mohamed. 2018. *How Somalia's Government Is Taking Big Strides to Stability*. June 26. https://www.centreforpublicimpact.org/somalia-government-stability/.

Legacy Center for Peace and Transparency. *Overview of Corruption, Underlying Causes, and Its Impacts in Somalia*. Maplewood, MN: Legacy Center for Peace and Transparency, 2016.

Lehrke, Jesse Paul, and Rahel Schomaker. "Kill, Capture, or Defend? the Effectiveness of Specific and General Counterterrorism Tactics against the Global Threats of the Post-9/11 Era." *Security Studies* 25, no. 4 (2016): 729–762. doi:10.1080/09636412.2016.1220199.

Macdonald, Julia, and Jacquelyn Schneider. "Presidential Risk Orientation and Force Employment Decisions: The Case of Unmanned Weaponry." *Journal of Conflict Resolution* 61, no. 3 (2017): 511–536. doi:10.1177/0022002715590874.

Mahmood, Faiqa. "The Efficacy of US Drone Strikes in Pakistan: The Long View." *Yale Journal of International Affairs* (2016). http//yalejournal.org/article_post

Marqaati. *Somalia State of Accountability 2016*, 1–15. Mogadishu: Marqaati, 2017. https://marqaati.org/wp-content/uploads/2017/02/Marqaati-2016-Corruption-Report.pdf.

Maruf, Harun. 2019. *Somalia Assesses Al-Shabab Moles' Infiltration of Government*. August 15. https://www.voanews.com/africa/somalia-assesses-al-shabab-moles-infiltration-government.

Mazhar, Ummad. "Do Remotely Piloted Aerial Vehicles Make Terrorism More Costly for Terrorists? Empirical Evidence from Pakistan." *International Journal of Conflict Management* 27, no. 4 (2016): 470–486. doi:10.1108/IJCMA-06-2015-0035.

McNeal, Gregory S. "Are Targeted Killings Unlawful? A Case Study in Empirical Claims without Empirical Evidence." In *Targeted Killings: Law and Morality in an Asymmetrical World*, edited by Jens David Ohlin, Andrew Altman, and Claire Finkelstein, 326–346. Oxford: Oxford University Press, 2012.

Menkhaus, Ken. "Managing Risk in Ungoverned Space: Local and International Actors in Somalia." *SAIS Review of International Affairs* 36, no.1 (2016): 109–120. Johns Hopkins University Press. doi:10.1353/sais.2016.0008.

Michel, Holland, and Dan Gettinger. "Drone Year in Review: 2017." Annandale-on-Hudson, NY: Center for the Study of the Drone at Bard College, 1–21. 2018. http//dronecenter.org

Ministry of Planning, Investment and Development. *Aid Flows in Somalia: Analysis of Aid Flow Data*. Mogadishu: Somalia Ministry of Planning, Investment and Economic Development, 2018.

Mogadishu-based Businessperson, Interview by Author. 2019. (May 7).

Mogadishu-based Official, Interview by Author. 2019. (April 4).

Morgan, Wesley. 2017. *U.S. Military Builds up in Land of 'Black Hawk Down' Disaster.* November 19. https://www.politico.com/story/2017/11/19/troops-somalia-military-buildup-247668.

Muia, Joseph. 2019. *Somali President Farmajo Renounces His U.S. Citizenship*. August 2. https://citizentv.co.ke/news/somali-president-farmajo-renounces-his-u-s-citizenship-267209/.

Nissenbaum, Dion. 2017. "U.S. Forces Get More Freedom to Strike Militants in Somalia." *The Wall Street Journal*, March 30. https://www.wsj.com/articles/u-s-forces-get-more-freedom-to-strike-militants-in-somalia-1490924860.

Pashakhanlou, Arash Heydarian. "Air Power in Humanitarian Intervention: Kosovo and Libya in Comparative Perspective." *Defence Studies* 18, no. 1 (2018): 39–57. doi:10.1080/14702436.2017.1420420.

Pratt, Simon Frankel. 2018. *US Killing by Drone: Continuity and Escalation*. December 11. https://www.lowyinstitute.org/the-interpreter/us-killing-drone-continuity-and-escalation.

Price, Bryan C. "Targeting Top Terrorists: How Leadership Decapitation Contributes to Counterterrorism." *International Security* 36, no. 4 (2012): 9–46. doi:10.1162/ISEC_a_00075.

Purkiss, Jessica, and Abigail Fielding-Smith. 2018. *Somalia: Reported US Actions 2018.* https://www.thebureauinvestigates.com/drone-war/data/somalia-reported-us-actions-2018#strike-27118.

Reeder, Bryce W., and John R. Smith. "US Strikes in Somalia and Targeted Civilian Killings by Al-Shabaab: An Empirical Investigation." *Foreign Policy Analysis* 15, no. 4 (2019): 589–603. doi:10.1093/fpa/orz008.

Renic, Neil C. "Justified Killing in an Age of Radically Asymmetric Warfare." *European Journal of International Relations* 25, no. 2 (2018): 408–430. doi:10.1177/1354066118786776.

Reno, William. "The Politics of Security Assistance in the Horn of Africa." *Defence Studies* 18, no. 4 (2018): 498–513. doi:10.1080/14702436.2018.1463819.

Reno, William. "The Dilemmas of Security Assistance to a Failed State: Lessons from Somalia." In *Before Military Intervention: Upstream Stabilisation in Theory and Practice*, edited by Timothy Clark and Robert Johnson, 53–77. London: Palgrave Macmillan, 2019.

Rinehart, Christine S. *Drones and Targeted Killing in the Middle East and Africa: An Appraisal of American Counterterrorism Policies*. Washington DC: Lexington Books, 2016.

Rossiter, Ash. "Drone Usage by Militant Groups: Exploring Variation in Adoption." *Defense & Security Analysis* 34, no. 2 (2018): 113–126. doi:10.1080/14751798.2018.1478183.

Rossiter, Ash, and Brendon J Cannon. "Re-examining The" Base": The Political and Security Dimensions of Turkey's Military Presence in Somalia." *Insight Turkey* 21, no. 1 (2019): 167–188. doi:10.25253/99.2019211.09.

Runkle, Benjamin. 2013. "The Lost Lessons of 'Black Hawk Down'." *War on the Rocks*. https://warontherocks.com/2013/10/the-lost-lessons-of-black-hawk-down/.

Samatar, Abdi Ismail. *Africa's First Democrats: Somalia's Aden A. Osman and Abdirazak H. Hussen*. Bloomington: Indiana University Press, 2016.

Samatar, Ahmed I. 2018. *Somaliland's Stakes in a Fast Changing Horn of Africa.* December 5. https://somalilandchronicle.com/2018/12/05/somalilands-stakes-in-a-fast-changing-horn-of-africa-professor-ahmed-i-samatar/.

Savage, Charlie. 2017. *Will Congress Ever Limit the Forever-Expanding 9/11 War?* October 28. https://www.nytimes.com/2017/10/28/us/politics/aumf-congress-niger.html.

Security Assistance Monitor. 2019. *Somalia.* https://securityassistance.org/data/program/military/Somalia/.

Shah, Aqil. "Do U.S. Drone Strikes Cause Blowback? Evidence from Pakistan and Beyond." *International Security* 42, no. 4 (2018): 47–84. doi:10.1162/isec_a_00312.

Sheikh, Abdi, and Feisal Omar. 2018. *Exclusive: Weapons Stolen from UAE Training Facility in Somalia, Sold on Open Market.* April 25. https://www.reuters.com/article/us-somalia-arms/weapons-stolen-from-uae-training-facility-in-somalia-sold-on-open-market-idUSKBN1HW26I.

Smith, Megan, and James Igoe Walsh. "Do Drone Strikes Degrade Al Qaeda? Evidence from Propaganda Output." *Terrorism and Political Violence* 25, no. 2 (2013): 311–327. doi:10.1080/09546553.2012.664011.

Sperber, Amanda. 2019. *Inside the Secretive US Air Campaign in Somalia.* February 7. https://www.thenation.com/article/somalia-secret-air-campaign/.

START, National Consortium for the Study of Terrorism and Responses to Terrorism. 2018. *Global Terrorism Database [Data File].* https://www.start.umd.edu/gtd.

Strawser, Bradley Jay. "Moral Predators: The Duty to Employ Uninhabited Aerial Vehicles." *Journal of Military Ethics* 9, no. 4 (2010): 342–368. doi:10.1080/15027570.2010.536403.

Transparency International. 2019. *2018 Corruptions Perception Index.* https://www.transparency.org/cpi2018.

Turse, Nick. 2019. *Pentagon Document Shows U.S. Knew of "Credible" Reports of Civilian Casualties after Its Attacks in Somalia.* July 10. https://theintercept.com/2019/07/10/pentagon-airstrikes-civilian-casualties-somalia/.

U.S. Africa Command Public Affairs. *U.S. Conducts Airstrike in Support of the Federal Government of Somalia.* Stuttgart: AFRICOM, 2018. https://www.africom.mil/media-room/pressrelease/31094/u-s-conducts-airstrike-in-support-of-the-federal-government-of-somalia.

U.S. Department of State. 2018. "Country Reports on Human Rights Practices: Somalia." Washington, DC. https://www.state.gov/reports/2018-country-reports-on-human-rights-practices/somalia/.

United Nations. *Report of the Monitoring Group on Somalia and Eritrea Pursuant to Security Council Resolution 2111 (2013): Somalia S/2014/726.* New York: UN General Assembly, 2014.

Vandiver, John. 2019. *US Helps Somali Commando Force Grow from Battalion to Brigade.* March 22. https://www.stripes.com/news/us-helps-somali-commando-force-grow-from-battalion-to-brigade-1.573794.

Vilmer, Jean-Baptiste Jeangène. 2017. "The French Turn to Armed Drones." *War on the Rocks*, September 22.

Vlassenroot, Koen, Sandrine Perrot, and Jeroen Cuvelier. "Doing Business Out of War. An Analysis of the UPDF's Presence in the Democratic Republic of Congo." *Journal of Eastern African Studies* 6, no. 1 (2012): 2–21. doi:10.1080/17531055.2012.664701.

Waltz, Kenneth N. *Theory of International Politics.* Long Grove, IL: Waveland Press, 2010.

Williams, Paul D. *Fighting for Peace in Somalia.* Oxford: Oxford University Press, 2018.

Williams, Paul D. 2017. "Paying for AMISOM: Are Politics and Bureaucracy Undermining the AU's Largest Peace Operation?" *IPI Global Observatory*, January 11. https://theglobalobservatory.org/2017/01/amisom-african-union-peacekeeping-financing/.

Williams, Paul D. "Joining AMISOM: Why Six African States Contributed Troops to the African Union Mission in Somalia." *Journal of Eastern African Studies* 12, no. 1 (2018): 172–192. doi:10.1080/17531055.2018.1418159.

Williams, Paul D. "Subduing al-Shabaab: The Somalia Model of Counterterrorism and Its Limits." *The Washington Quarterly* 41, no. 2 (2018): 95–111. doi:10.1080/0163660X.2018.1484227.

Wilner, Alex. "Targeted Killings in Afghanistan: Measuring Coercion and Deterrence in Counterterrorism and Counterinsurgency." *Studies in Conflict & Terrorism* 33, no. 4 (2010): 307–329. doi:10.1080/10576100903582543.

Xinhua. 2019. "AMISOM Commits to Peaceful 2020/21 Polls in Somalia." *Xinhua*. November 30. http://www.xinhuanet.com/english/2019-11/30/c_138595814.htm.

What's wrong with drones? Automatization and target selection

Andree-Anne Melancon

ABSTRACT
Developments in artificial intelligence have ignited many debates on the use of autonomous weapons and 'killer robots.' However, before autonomous attack decisions can be explored, issues related to target selection need to be considered. Focusing the discussion on drones and artificial intelligence overshadows more fundamental issues. This paper therefore argues that the main ethical problem raised by autonomous drones do not come from the weapon system or technology itself. Instead, they stem from challenges of target identification. Thus, asking 'what's wrong with drones?' is misleading as it focusses on the technology instead of the primary issue of target selection.

Introduction

The development of lethal autonomous weapon systems (LAWS) has raised many concerns. In December 2016, at the Fifth Review Conference of the High Contracting Parties to the Convention on Prohibitions or Restrictions on the Use of Certain Conventional Weapons Which May be Deemed to be Excessively Injurious or to Have Indiscriminate Effects (commonly known as the CCW convention) it was decided that an 'open-ended Group of Governmental Experts (GGE) related to emerging technologies in the area of lethal autonomous weapons systems'[1] was necessary to monitor the development of these weapons. This is indicative of a certain unease in the international community with the development of LAWS. Moreover, in March 2018, a group of over 50 researchers and engineers signed an open letter to Professor Sung-Chul Shin, president of the Korea Advance Institute of Science and Technology (KAIST), declaring their boycott of the institution following the announcement it was partnering with Hanwha Systems, South

Korea's largest defence contractor.[2] The signatories expressed concerns that the partnership would accelerate the race to develop autonomous lethal weapons, which led Sung-Chul Shin to respond stating that KAIST will not develop 'killer robots.'[3] More broadly, there is a growing sense of urgency regarding the potential regulation of LAWS since technology cannot be 'un-invented' and their development is seen as a Pandora's box that will be difficult to close. Groups such as the Campaign to Stop Killer Robots have called for a complete ban of the 'development, production and use of fully autonomous weapons.'[4] Importantly, the debates surrounding the development of LAWS generally focus on issues linked to the weapons themselves and the consequences of the technology falling into the hands of 'despots and terrorists,'[5] ignoring existing issues with the implementation of certain principles of the law of armed conflict in contemporary contexts.

Indeed, as will be argued in this paper, the identification of legitimate targets remains an important challenge that is too often overlooked. In addition to being important in itself, target identification is also key since it underpins all use of force (it is the first step in determining if/how force is used). Thus, this paper argues that before issues related to automatization and artificial intelligence can be fully explored, attention should be given to the criteria upon which the use of force is based. To support this argument, this paper will review key debates on the use of LAWS to argue that the main ethical problems related to the use of automated lethal drones are not linked to technology, artificial intelligence, or the weapon systems themselves, but rather to the criteria on which the use of force is based.

Before the discussion on LAWS can be taken any further, it is important to clarify the difference between automated systems and autonomous systems. The two are similar and can easily be interchanged, but they refer to different types of systems and should not be conflated. *Automated* weapon systems operate following rules and clear structures. They use 'if-then' patterns, meaning that a given input will always generate the same output (if the system operates without mistakes or errors). By contrast, an *autonomous* system will not always generate the same output for a given input because the system operates on probability 'meaning that it makes guesses about best possible courses of action given sensor data input.'[6] In autonomous systems, the input is fed into a 'world model' built by the system that is dynamically evolving as it interacts with the environment. In other words, autonomous systems are constantly updated via machine learning. As explained by Missy Cummings, Director of the Humans and Autonomy Laboratory at Duke University:

> This means that the world must be perceived (or sensed through cameras, microphones and/or tactile sensors) and then reconstructed in such a way that the computer "brain" has an effective and updated model of the world it is in before it can make decisions. The fidelity of the world model and the timeliness of its updates are the keys to an effective autonomous system.[7]

In short, automated system follow strict rules and formulas producing uniform outputs whereas autonomous systems produce a variety of outputs.

The recent debates and publicity surrounding the development of driverless cars has brought fully autonomous vehicles into the public's eye, especially following a 2018 incident where a woman was killed by an autonomous Uber in Arizona 'in what appears to be the first reported fatal crash involving a self-driving vehicle and a pedestrian in the US.'[8] These types of driverless vehicles operate following maps whilst also developing an understanding of all other moving people and things nearby (pedestrians, cyclists, other vehicles, animals, etc.). Driverless cars ought to build complex dynamic world models accounting for aspects of their environment. They need to assess the dynamics of all other vehicles and obstacles to identify possible intersections, predicting the way traffic will behave to determine how to act.

> This form of estimating or guessing what other drivers will do is a key component of how humans drive, but humans do this with little cognitive effort. It takes a computer significant computation power to keep track of all these variables while also trying to maintain and update its current world model.[9]

It is important to note that in comparison, the world model developed by an autonomous unmanned aircraft is less complex since it includes fewer imponderables, focussing on maps, preferred routes, no-fly zones, and other obstacles.[10] Therefore, since the world model required by aircrafts is less advanced than the one required by a driverless car, it is not inconceivable that fully automated and autonomous unmanned aircrafts become a feature of conflict before driverless cars take over the roads or unmanned ground vehicles are deployed on battlefields.[11]

Key debates

Much like with the development of driverless cars, advances in the field of autonomous aviation generate many debates, especially when it has to do with lethal automated weapons which could make decisions on the use of force.[12] This section will focus on two key debates that have surrounded the development of LAWS. Firstly, issues surrounding the nature of the human-machine interface will be explored. Following this, issues and concerns linked to human dignity will be reviewed. Through these debates, it will be highlighted that although valid points are raised, they overshadow fundamental and pressing issues linked to target identification in contemporary conflict.

Human-machine interface

In the aftermath of the CCW expert meeting of 2018, Amnesty International issued a press release stating that 'the majority of states at the last CCW meeting in April 2018 emphasized the importance of retaining human control over weapons systems and the use of force, and expressed support for developing new international law on lethal autonomous weapons systems.'[13] Indeed, one of the most discussed concerns with the development of autonomous weapons has to do with artificial intelligence and the degree of human oversight that will remain involved in the use of force. The GGE 2018 meeting report reiterated that the nature of the human-machine interface ought to be considered at all stages of the development and deployment of LAWS:

> Importantly, In line with the Chair's 'sunrise slide', the following touch points in the human-machine interface were considered: (0) political direction in the predevelopment phase; (1) research and development; (2) testing, evaluation and certification; (3) deployment, training, command and control; (4) use and abort; (5) post-use assessment.[14]

Furthermore, the GGE argue that 'accountability threads together these various human-machine touch points in the context of the CCW. Humans must at all times remain accountable in accordance with applicable international law for decisions on the use of force.'[15]

Concerns and issues related to accountability and the level of human oversight or supervision can be divided in two categories: technical questions and ethical or policy considerations. The former refers to whether the system *can* make accurate decisions (on target identification for example), whereas the latter explore whether autonomous systems *should* make the decision to use lethal force. To fully understand the debates on artificial intelligence and automatization, it is important to understand reasoning and decision-making. Cummings[16] effectively builds on the framework developed by Rasmussen[17] to explain the relationship between reasoning and certainty. She identifies four types of behaviours with an increasing level of complexity: skill-based, rules-based, knowledge-based, and expertise-based.[18]

Firstly, skill-based behaviours become automatic for humans after a certain amount of training since they rely on the perception-action cycle (undertaking 'if-then' actions). Equation-based programming can be used to allow *automated* machines to carry out skills-based behaviours where the machine is programmed to always produce output B in response to input A.

Secondly, increasing the complexity of the tasks, rules-based reasoning relies on rules and procedures such as checklists. As Cummings elaborates:

> For instance, when a fire-light illuminates or another subsystem indicates a problem, pilots are trained to first stabilize the aircraft (a skill) but then turn to the manual to determine the correct procedure (rule following). Such codified procedures are necessary since there are far too many solutions to possible problems to be remembered.[19]

The US Air Force has reported that if a Global Hawk (RQ-4) drone loses communications, it enters a loitering pattern before it lands itself. This is an example of an automated rules-based behaviour. Yet, the development of driverless cars requires a further move beyond automation towards autonomy. Although driving (and navigation more broadly) is mainly a rules-based behaviour requiring a knowledge of traffic rules, it also required the ability to interpret and react to uncertainty.

This brings into play the third tier of Cumming's framework which is knowledge-based behaviours. As the spectrum moves from skills towards expertise, the ability to deal with uncertainty becomes increasingly necessary. Here, uncertainty reaches levels where algorithms cannot be used to understand and interpret situations since it is difficult to recognise what rules apply. 'It is,' Cummings posits, 'in these uncertain scenarios – which are by definition vague and ambiguous – that algorithms may not be able to understand the solution space, much less achieve a feasible solution.'[20] This is where knowledge-based behaviours take place (when situations do not match the established rules). The growing importance of interpretation of complex situations brings the move from automation to *autonomy*. The 'miracle on the Hudson' where US Airways flight 1549 was successfully ditched in the Hudson river in January 2009 after a double engine failure[21] is an example of a situation where the pilot carried out knowledge-based reasoning drawing on experience and rapid mental calculations.

Finally, Cumming's typology culminates with expertise-based behaviours. These build on knowledge-based behaviours and rely on a mix of expertise, intuition, and rapid assessments.[22] Experts have the ability to make challenging decisions rapidly, even in the face of uncertainty and ambiguity, without needing to take up time comparing all possible alternative courses of action. After all, in humans, the hallmark of expertise is the ability to come with highest levels of uncertainty and such behaviour remains difficult to replicate for computers. This is because a core aspect of induction is the ability to generalise in order to overcome uncertainty and ambiguity, which cannot be achieved with algorithms. Indeed, as Cummings argues: 'replicating the intangible concept of intuition, knowledge-based reasoning and true expertise is, for now, beyond the realm of computers.'[23]

When discussing the development of artificial intelligence, it is important not to conflate a computer's ability to execute specific tasks set or programmed by humans with independent knowledge. Although machine learning is moving towards the latter, humans remain currently involved to various degrees in the decision to use of force.

With the development of autonomous unmanned aircrafts, two key tasks are undertaken: navigation and weapon release. In regard to the former, and as discussed earlier, navigation is a relatively straightforward rules-based behaviour. Since there is less uncertainty in aerial transportation compared to ground transport (in terms of limited traffic and potentially unexpected encounters), autonomous aerial systems could be developed without raising too many concerns, especially when considering that much of commercial aviation already relies on automation algorithms such as autopilots. However, the second task is much more complex. Indeed, ambiguity and uncertainty are high when it comes to the use of force and weapon release, bringing this task in the realm of expertise-based behaviours. Although there is some automation, such as in the use of precision-guided missiles which are programmed to reach a specific target, the decision to launch such missile is currently made by humans. The key issue therefore has to do with autonomous of weapon release, which relies on knowledge and expertise.

Exploring autonomous weapon release raises questions about human oversight and accountability in the use of force. In 2018, the GGE suggested human responsibility as a guiding principle for the development of LAWS: 'Human responsibility for decisions on the use of weapons systems must be retained since accountability cannot be transferred to machines. This should be considered across the entire life cycle of the weapons system.'[24]

Following the GGE's six key moments for human involvement in the use of LAWS presented earlier,[25] weapon release takes place in the 'use and abort' phase. The GGE suggests that human oversight ought to remain at all times: 'Humans must at all times remain accountable in accordance with applicable international law for decisions on the use of force.'[26] However, it is unclear how this can be reconciled with autonomous weapons where machines are making decisions. As Amanda Sharkey argues, an autonomous 'weapon or robot cannot be held responsible for its actions.'[27] Therefore, the problem of who can be held accountable should there be any issues with the use of force will need to be resolved. However, since the technology is still being developed, it is difficult to make decisions on the format of the human-machine interface. Importantly, the way the machine makes the decision to use force will also need to be determined before the role of humans in the targeting cycle can be defined.

In short, the nature of the human-machine interface in the context of automated weapons remains unclear and problematic. Indeed, the non-neglectable questions of responsibility and accountability need to be solved before autonomous weapons can be deployed. The role and importance that humans will retain on the use of force will need to be defined. Options include 'human-*on*-the-loop weapons' and 'human-*out*-of-the-loop weapons.'[28] In both cases, the machine selects the targets and uses force against it but the former retains human oversight (and therefore the possibility to override the robot's actions).

Broadly, issues of responsibility and accountability in the use of LAWS refer to the targeting and engagement cycle (decision making leading to weapons release). This debate focusses on *who* is responsible for the decision to use force, ignoring another key aspect of the use of force: the targeting criteria upon which the decision to use force is based. Indeed, the debates on the human-machine interface have a tendency to overshadow issues related to target selection, which need to be addressed before the human-machine interface can be defined as it will have an impact on the type and format of human oversight deemed necessary.

Human dignity

The second area of debate linked to the development of LAWS has to do with their impact on human dignity. Although there are a variety of meanings associated with the terms, much of the argumentations against the use of LAWS generally builds on Kant's account of human dignity as being a special status (higher norm).[29] Extremely briefly, Kant views humans as having dignity, making them an end in themselves whereas objects are means that have a price and that can be used.[30] Therefore, 'death by algorithm' through the use of autonomous lethal weapons is seen as being opposed to human dignity because the human targets are treated as objects:

> to give a programmed machine the ability to "decide" to kill a human is to abandon the concept of human dignity. Humans are sometimes accidentally killed by machines, but for an autonomous robot/drone to be programmed to kill a human is to treat a rational being as if it were merely an object.[31]

Similarly, Christof Heyns, former UN Special Rapporteur on Extrajudicial, Summary or Arbitrary Executions, builds on Kant's understanding of dignity to argue that giving robots 'the power of life and death' is demeaning and an affront to human dignity because humans are reduced to objects. In 2014, Heyns presented a report on the promotion and protection of human rights to the UN General Assembly which raised concerns that the increased use of unmanned systems would lead to a depersonalization of force.[32] Similar concern had previously been raised by Heyns's predecessor, Philip Alston, with the disputable notion of the 'PlayStation Mentality' where it is thought that drone operators see their targets as mere 'blips' on a screen because they grew up playing videogames:

> Young military personnel raised on a diet of video games now kill real people remotely using joysticks. Far removed from the human consequences of their actions, how will this generation of fighters value the right to life? How will commanders and policy makers keep themselves from the deceptively antiseptic nature of drone killings?[33]

The 2014 report takes the discussion further and raises concerns for the right to life and dignity in response to the use of unmanned and autonomous systems by law enforcement agencies.[34] It questioned whether lethal autonomous weapons would be able to determine the permissible level of force considering the subtleties of human behaviour. Moreover, it stated that using autonomous weapons would be 'likely to be seen in many contexts as adding insult to injury, and an affront to human dignity.'[35] Although this raises interesting points, it also sets unrealistic expectations for autonomous machines when they are required to understand intentions. Since intentions are deeply personal and private, they cannot easily be judged or assessed by external forces, be it human or machine.

The report also raises concerns regarding the level of force used since automated or autonomous weapons are limited to the use of lethal force, without the possibility for capture. Yet, this concern requires a clarification: nearly any use of force that is not a 'boots on the ground' strike operation excludes the possibility for capture. Indeed, snipers, aircrafts (both manned and unmanned), missiles, tanks, etc., are limited to the use of lethal force insofar as they cannot capture enemy fighters. Moreover, there is no legal requirement to offer the opportunity to surrender before launching a strike; this requirement can exist in law enforcement but not under the law of armed conflict. Building on these concerns for human dignity, Human Right Watch states on its website that LAWS undermines human dignity because inanimate machines cannot comprehend the callus of individual life nor the significance of its loss: 'Allowing them to make determinations to take life away would thus conflict with the principle of dignity.'[36]

The argument that humans are treated as objects by autonomous weapons can be refined to state that: 'killing a human being in the hope that it will prevent further harm is insufficiently morally grounded to override human dignity.'[37] This is because humans are seen as objects that are sacrificed. However, this argument assumes the innocence of the targets, such as the passengers of a highjacked aircraft that would be shot down to prevent it from causing greater harm. This position rejects the possibility of dignity existing in the sacrifice of the few to save the many (or even only to save others). This argument also ignores the possibility that the individuals being targeted have done something to make them legitimate targets in the conflict where force is being used.

Moreover, the use of LAWS is seen as an attack on human dignity because there is no due process. This argument draws a distinction between the lawful killing of enemy combatants in conflict, which does not require judgement, and the killing of individuals as punishment with autonomous weapons.[38] Here again, the target is seen as a means to an end deprived of human dignity. However, even the killing of enemy combatants can be seen as some form of punishment since they are targeted for some reason, such as fighting on the 'wrong side' (defined in terms of a wider political and military strategy such as 'terrorism' in many cases) for example. In conflict, killing is differentiated from gratuitous violence or irrational behaviour because the use of lethal force is justified (within certain limits). These justifications identify the target as being liable to have force used against them either because of membership to the enemy's army or because of certain actions being carried out. Michael Walzer argues that the first war convention implies that all soldiers are equal in war and that 'simply by fighting, whatever their private hopes and intentions, they have lost their title to life and liberty.'[39] This means that all killing can be considered as being a certain form of punishment and therefore it is unclear how the killing done via an autonomous weapon would constitute a form of punishment whereas the use of non-autonomous aircrafts or snipers would be lawful killing. In a Lawfare blog post, Adam Saxton has effectively argued that the human dignity is impacted in warfare as a whole, not particularly by the use of autonomous weapons. Considering this, the distinction between lawful killing carried out in conflict and the perceived 'punishment killings' carried out by autonomous weapons becomes blurred. Further, this discussion on due process in the use of force indicates that the way targets are selected is of great importance, yet this issue is not explored further. This is in line with the central argument of this paper that the more fundamental issue with the development of automated weapons does not regard the technology or artificial intelligence. Instead, the issues stem from the way targets are selected.

Additionally, debates on human dignity raise a wider question on whether the type of weapon doing the killing has an impact. Heyns's 2014 report states that the use of automated weapons adds 'insult to injury' which indicates a belief it does. Indeed, proponents of the human dignity opposition argue that it does because 'the perceived benefits of autonomous weapons are one-sided without any relation to the individual target.'[40] However, a review of the human dignity argument indicates that much of the opposition to autonomous weapons comes from the machine making the *decision* to kill someone, rather than with the machine using force. In other words, issues come from the autonomous identification of legitimate targets (where machines make life or death decisions) rather than with the autonomous technology as a whole. Although this argument is concerned with target selection, it ignores *how* targets are identified and simply focusses its concerns on who, or rather *what*, is making that determination. Consequently, these discussions are incomplete since they overlook the criteria leading to target identification, a non-neglectable aspect of the use of force.

Finally, Ulgen has argued that the use of LAWS not only undermines the human dignity of the target but it also undermines the dignity of combatants deploying such weapons 'by not recognizing their professional training and military ethics of courage and respect for human targets.'[41] This statement highlights a certain unease surrounding distance killing. Indeed, the development of unmanned aircrafts generated debates on the 'fairness' of their use because of the risk imbalance between the operators and the targets on the ground.[42] Alex Leveringhaus suggested in an ICRC [International Committee of the Red Cross] blog that issues related to the increasing distance in fighting require 'introspection, if not a change in perspective.' Furthermore, Christian Enemark argues that the use of unmanned weapons generates situations of 'radical asymmetry' where 'an imbalance [is] so obvious and so profound that one side is apparently unable to apply – in a just fashion, at least – any strength at all against the other.'[43] Moreover, Ulgen argues that 'autonomous weapons create a hierarchy of human dignity whereby human targets are unequal and unworthy of face-to-face human contact and therefore excluded from an inclusive moral theory.'[44] However, these issues of fairness and risk balance are valid and applicable to all distance use of force, whether manned or not. Soldiers have always sought to create as much distance between themselves and their enemy, with the use of slingshots, bows and arrows, catapults, missiles, etc. If fighting is not taking place in face-to-face, reciprocal conditions resembling a boxing match, then the use of force is unequal. Yet, inequality does not automatically create a hierarchy of dignity. Rather, inequality is a constant feature of warfare and fighting. Conflicts where all sides are roughly equal bring periods of stalemate where the conflict is 'frozen' until one side can outweigh its adversaries. Thus, concerns about inequality must bear in mind that warfare is not about equality, it is not a game of chess with clearly defines rules, parameters, and equal opportunities. Unmanned aircrafts are a fairly new form of military technology whose use is meeting strong moral aversion. This type of initial aversion is often seen with the development of new and unprecedented forms of fighting such as long-range missiles or archery. Yet, initial antipathy does not mean that the weapons are inherently problematic.

Additionally, regarding the use of unmanned aircrafts, Bradley J. Strawser has argued that there is a moral duty to employ unmanned systems if they are available following the principle of unnecessary risk.[45] Furthermore, this is also be echoed by the principle of humanity, which is one of the four fundamental principles of the law of armed conflict (LOAC). Indeed, this principle prohibits unnecessary suffering. Importantly, LOAC exists to protect both combatants and non-combatants and the principle of humanity is the reason behind the ban on many means of warfare such as flamethrowers and laser-blinding weapons for example. Thus, the principle of humanity could support the deployment of unmanned aircrafts as it would prevent some unnecessary suffering, as suggested by the principle of unnecessary risk. Often, human dignity is presented as being a principle of *humanity*.[46] However, is it important to distinguish this understanding from the law of armed conflict's principle of humanity presented above. Human dignity as humanity refers to 'human-ness' and not to the LOAC principle. Indeed, arguments on human dignity oppose the use of autonomous weapons because they treat people as objects (as less-than human).

Ulgen takes her opposition to autonomous weapons further and argues that their deployment would mean that 'there would also appear to be no reason for having armies.'[47] This is a vast oversimplification and assumes that the only role fulfilled by militaries is to eliminate enemy targets. Yet armies and militaries undertake a variety of roles including humanitarian assistance, capacity building, and disaster relief, in addition to conventional warfighting. Assuming that the development of autonomous weapons will make armies unnecessary is to ignore the wide scope of roles militaries play at home and abroad.

The main strength of the human dignity line of argumentation is that it appeals to emotions, which can lead to greater opposition to the use of LAWS. To that effect, Amanda Sharkey has suggested that a key campaigning advantage of this argumentation is that it evokes 'a strong visceral response.'[48] Yet, arguments grounded on human dignity remain unconvincing. It is unclear why autonomous weapons should be banned and not warfare more broadly. As Ariadna Pop rightly observed in an ICRC blog post, it remains unclear *how* human dignity is violated by the use of LAWS: 'Unless it can be shown why the concept of human dignity mandates the prohibition of [L]AWS, such a prohibition is not justified.'[49] Sharkey comes to a similar conclusion after having assessed the human dignity arguments against the use of LAWS and argues that 'although dignity is an important concern when considering LAWS, it is not sufficient on its own.'[50] Thus, it can be argued that much of the focus on human dignity has overshadowed more fundamental problems related to the use of force more broadly. One of those greater problem is target identification in contemporary conflict. Its importance will be explored in the following section.

Target identification

As presented earlier, much of the debates surrounding the development of autonomous drones focus on the human-machine interface ignoring the more fundamental issue of target identification. Similarly, argumentations around human dignity focus on individuals being treated as objects without exploring why certain individuals are targeted. Yet, for the use of force to be considered morally just and lawful, it ought to respect certain criteria. As Michael Walzer remarks: 'war is always judged twice, first with reference to the reasons states have for fighting, secondly with reference to the means they adopt.'[51] In the context of exploring the use of LAWS, it is this latter assessment focussing on *jus in bello* that is of importance. This set of criteria judges the way war is fought and is generally conceived as having two central criteria: distinction and proportionality. Distinction sets 'how those victims of war who can be attacked and killed are to be distinguished from those who cannot'[52] whereas proportionality balances the benefits and consequences of the use of force.

Too often, the newness of technology overshadows more fundamental issues in the use of force. New weapons are developed but the same 'old' problems remain. Key matters to be explored and assessed in the use of force can be summarised to the law of armed conflict's four basic principles of distinction, proportionality, military necessity, and humanity. Though all these principles are essential and important, distinction often underpins the implementation of the others – the proportionality calculations of the use of force will depend on who or what is the intended target for example. With the development of autonomous drones, debates on the human-machine interface and human dignity have overshadowed the more fundamental issue of target identification. Albeit the nature of the human-machine interface will indeed need to be defined and questions of accountability will need to be answered, they are secondary to the programming and functioning of the autonomous weapons, including the criteria on which the use of force is based. As argued earlier, objections to autonomous weapons on grounds of human dignity focus on the objectification of targeted individuals, without exploring *why* certain individuals are identified as legitimate targets in the first place.

International law, and therefore target identification to a certain extent, was explored during the 2018 CCE GGE meeting. Alongside discussions on the categorisation of LAWS, the human element in the use of force, and military applications of technology, the GGE explored the humanitarian and security challenges brought by LAWS.[53] The discussions ultimately stated that the law of armed conflict applies to all weapons, including autonomous ones[54] and therefore 'as IHL [international humanitarian law] is fully applicable to potential lethal autonomous weapons systems a view was also expressed that no further legal measures were needed.'[55] However, this is not sufficient since it assumes that there are no issues with the way legitimate targets are identified in contemporary conflict, which is not the case.

Revising target selection

This final section argues that target identification is the most important challenge for the development and future use of automated weapons since it underpins the use of force. The main challenge here is that the move towards automatization and autonomous target identification requires clear criteria that can be programmed in order to allow for machines to identify legitimate targets autonomously. Target identification refers to the principle of distinction (or discrimination in the just war literature). This principle calls for the distinction between legitimate and illegitimate targets in conflict, where the latter are protected and the former can be targeted (in accordance with the other principles of the law of armed conflict). This requirement is of such importance that the ICRC identifies distinction as its first rule of international humanitarian law: 'Rule 1. The parties to the conflict must at all times distinguish between civilians and combatants. Attacks may only be directed against combatants. Attacks must not be directed against civilians.'[56] In addition to this, Rule 3 clarifies who is be considered a 'combatant': 'All members of the armed forces of a party to the conflict are combatants, except medical and religious personnel.'[57] This effectively summarises the legal definition of combatant found in the Geneva Conventions and their Additional Protocols. This understanding of distinction is applicable in the context of a conventional war only involving states and their armies. Looking at the automatization of the use of force, the uniforms of enemy armies could be codified and programmed into unmanned weapon system. This could allow the autonomous weapons to identify enemy fighters in order to launch strikes. Moreover, the tactical recognition flash or unit colour patch or shoulder sleeve insignia of the opposing armies' religious corps could also be coded and programmed to allow for the autonomous weapon system to distinguish protected religious personnel from legitimate targets. Likewise, the distinctive emblems of the Red Cross, Crescent, and Crystal could also be coded and programmed to allow the autonomous weapons to identify protected persons and places.

Yet, even in conventional inter-state conflict, the principle of distinction is more complex than simply identifying enemy soldiers. This is because soldiers are not always legitimate targets. Indeed, combatants can no longer be targeted when they are 'hors de combat' as a result of an injury or capture. Thus, autonomous weapons will need to be able to understand the difficult issue of 'hors de combat' combatants in order to distinguish between combatants that are still 'in combat' and those who are 'out of combat.' Another challenge for autonomous target identification is that some legitimate targets become protected while they carry out medical duties (and are considered as auxiliary medical personnel). Such actions include carrying a stretcher or providing first aid for example. These types of actions would also need to be coded to allow autonomous machines to understand that individuals carrying out auxiliary medical roles are protected during those duties. Further, autonomous machines will also need to be able to distinguish between genuine auxiliary roles and those who are involved in the conflict and carry a stretcher with them as an attempt to shield attacks. These two additional challenges emphasise the importance of inductive reasoning and expertise-based behaviours in target selection.

In regard to non-combatants, a simple definition of this category would include any individual who is not considered a combatant. Non-combatants are generally civilians but there is a limit to this: civilians lose their protection if they take a direct part in the hostilities. Indeed, Article 51 of the first Additional Protocol states: 'Civilians shall enjoy the protection afforded by this Section, unless and for such time as they take a direct part in hostilities.'[58] The notion of 'direct participation in hostilities' (DPH) exists as a limit to civilian immunity. Even considering its importance, there is no universally agreed upon definition of 'direct participation in hostilities' and the types of actions that constitute which DPH remain a topic of debate in both legal and philosophical circles.[59] In an attempt to overcome this, the ICRC organised a series of expert meetings between 2003 and 2008 to clarify the meaning and scope of DPH, but the process was unsuccessful.[60] Nevertheless, in 2009, the ICRC published its interpretative guidance on the notion of DPH, which is defined as referring to 'specific acts carried out by individuals as part of the conduct of hostilities between parties to an armed conflict.'[61] The ICRC guidance sets three constitutive elements that form DPH: a threshold of harm, a causal link, and a belligerent nexus.[62] These constitutive elements could set the foundation for a conflict-specific codification to be programmed into the autonomous weapons in order to serve as the foundation for the expertise-based decisions to use force.

Identifying legitimate targets in conventional inter-state conflict is complex but not impossible. The principle of distinction could potentially be coded to allow automatization and the use of autonomous weapons, if the issues of 'hors de combat' combatants, auxiliary medical personnel, and direct participation in hostilities can be adequately defined and programmed. However, the majority of contemporary conflicts are asymmetrical or non-international. This means that they involve state actor(s) fighting non-state actor(s).[63] In such situations, the legal definitions of combatants (associated with combatant status and prisoner of war status) do not apply. It is here that the challenge of identifying legitimate targets is the greatest.

> In some countries where targeted killings are carried out, potential victims may change "costumes". One day they are combatants, while the next day they may blend into the civilian population. They usually do not wear uniforms, which was the technique used in past war conflicts as an identification sign for combatants. Also, whether or not person is armed is not necessarily the distinguishing factors between civilian and non-uniformed combatants.[64]

Considering this, the principle of distinction is difficult to apply in situations of non-international armed conflict because of the difficulty to identify legitimate targets.

One way to overcome this would be to use the notion of direct participation in hostilities as a guiding principle, meaning that all individuals are considered illegitimate targets unless they take direct part in hostilities. This would require much work to clarify the scope and meaning of DPH in specific contexts. Actions that constitute DPH would need to be identified and coded in a way that would eliminate the risk of the autonomous weapon erroneously targeting innocent individuals. These DPH actions would set the foundations for expertise-based behaviours carried out by machines capable of inductive reasoning.

The idea of establishing lists of behaviours or patterns to identify legitimate targets in conflict is not new. For example, former U.S. Secretary of State Colin Powell stated that some profiling was used in Vietnam to identify targets: Helicopters encountering suspicious peasants wearing black 'pyjamas' would fire in front of the suspect in order to see the reaction, movement was seen as 'evidence of hostile intent' meaning that they were legitimate targets.[65]

More recently, patterns and profiles have been used in drone strikes carried out by the U.S. Indeed, in the absence of clearly defined combatants, patterns of life or behaviour are used as a means to identify legitimate targets and to launch what has been called 'signature strikes.'[66] These patterns or 'signatures' are identified as being indicative of combatant status, meaning that any individual who fits one of the patterns or signatures is acting like a combatant, and therefore can be considered a combatant, and therefore legitimate target. To a certain extent, this can be seen as an attempt to codify the notion of direct participation in hostilities. On paper, this appears rather straightforward and suggest that target identification could be coded and ultimately automated. However, the realities of conflict are too complex. Indeed, a brief look at the types of signatures used to legitimise the use of force indicates that target identification remains a key issue in contemporary conflict. Although there is no list of official signatures, research by Kevin Jon Heller in 2013 identified 14 signatures that are used to justify the use of force by the U.S. in the War on Terror:

(1) Planning attacks;
(2) Transporting weapons;
(3) Handling explosives;
(4) Al-Qaeda compound;
(5) Al-Qaeda training camp;
(6) Military-age men in area of known terrorist activity;
(7) Consorting with known militants;
(8) Armed men traveling in trucks in al-Qaeda in the Arabian Peninsula-controlled areas;
(9) Suspicious camp in al-Qaeda-controlled area;
(10) Groups of armed men travelling towards conflict;
(11) Operating an al-Qaeda training camp;
(12) Training to join al-Qaeda;
(13) Facilitators;
(14) Rest areas.[67]

Additionally, 'funerals' should also be included in this list since a pattern has been established where the funerals of enemy fighters are targeted.

This need to rely on signature strikes based on patterns of life or behaviour is indicative of the difficulties associated with the identification of legitimate targets in contemporary conflict. Importantly, this use of signatures is an inadequate way to implement the distinction principle which leads to the targeting of illegitimate targets. To that effect, Kristina Benson argues that signature strikes do 'a poor job of distinguishing between combatants and non-combatants, resulting in a disproportionate amount of civilian casualties.'[68] This therefore suggests that although the list of patterns of signatures might indeed include actions that can be indicative of enemy fighter status, they also encompass many actions that are undertaken by innocent individuals. For example, looking at signature six from the list above, this can easily lead to a great number of individuals being considered enemy fighter without any additional proof of their actual involvement in the hostilities. Thus, when looking at the future autonomous targeting, the machines will need to move beyond rules-based 'if-then' behaviours in order to avoided targeting any slightly suspicious individuals. Targeting cannot simply be automated (rules-based), it will be autonomous since it requires expertise-based behaviours. Indeed, autonomous machines will need to be able to analyse complex situations where certain listed patterns can be rapidly interpreted with expertise and intuition. Nevertheless, even if the identification of targets is moving away from rules-based behaviours, it will remain founded on certain criteria which need to be effectively defined. Thus, greater focus and attention is required on the issue of target identification before further questions related to the autonomous use of force can be addressed fully. It is important not to let the newness of the emerging technology overshadow the fundamental and lasting issue of target identification.

Conclusion

This paper has explored some of the key ethical issues raised by the development of unmanned and autonomous drones. The nature of the human-machine interface will indeed need to be determined and the level of human oversight over the autonomous actions still needs definition. However, it is difficult to make these determinations without knowing how the future machines will functions and how they will be used. This last point brings into question the issue of target identification, which underpins the deployment of LAWS. Another concern with the use of autonomous weapons has to do with human dignity. This line of argumentation raises many relevant points, but it remains unclear how the deployment of LAWS is a unique affront to human dignity. Nevertheless, the review of the human dignity line of argumentation highlights that much of the concerns raised are relevant for all uses of force, be it autonomous or not, especially when it discusses treating humans as objects. This then leads to the central argument of this paper that the main ethical problems with the use of autonomous drones do not come from the technology or the weapon systems themselves. Rather, much of the issues are rooted deeper and regard the way targets are identified in contemporary conflict.

In closing, asking 'what's wrong with drones?' is misleading since it focusses on the weapon system and the technology instead of exploring the primary issues of target selection. Indeed, target identification will underpin the use of lethal autonomous weapon systems since these vehicles will undertake two main activities: navigation (including intelligence gathering) and weapon release. Therefore, it is essential to have a clear understanding of how legitimate targets are identified in every context where force might be used. Even with future LAWS undertaking expertise-based behaviours in the selection of their targets, these behaviours will remain founded on some knowledge and rules. These rules implement the distinction principle by identifying the characteristics of legitimate targets. In order to achieve this, the concept of direct participation in hostilities will need to be refined and clearly defined. There is nothing fundamentally wrong with drones as weapons. The development of this new technology has simply highlighted existing issues in contemporary *jus in bello*. Instead of focussing on the technology itself or exploring future issues of accountability, the more pressing challenge of implementing the principle of distinction in contemporary conflict need to be addressed.

Notes

1. Meeting of the High Contracting Parties to the Convention on Prohibitions or Restrictions on the Use of Certain Conventional Weapons Which May Be Deemed to Be Excessively Injurious or to Have Indiscriminate Effects, Final Report.
2. The full text of the "Open Letter to Professor Sung-Chul Shin, President of KAIST from some leading AI researchers in 30 different countries" is available online: https://www.cse.unsw.edu.au/~tw/ciair//kaist.html.
3. His response stated: 'As an academic institution, we value human rights and ethical standards to a very high degree, I reaffirm once again that KAIST will not conduct any research activities counter to human dignity including autonomous weapons lacking meaningful human control.' Cited in Haas, "Killer Robots".
4. Campaign to Stop Killer Robots, "The Problem."
5. "Open Letter to Professor Sung-Chul Shin."
6. Cummings, *Artificial Intelligence and the Future of Warfare*, 3.
7. Ibid., 4.
8. Levin and Wong, "Self-Driving Uber Kills Arizona Woman in First Fatal Crash Involving Pedestrian."
9. Ibid.
10. Ibid.
11. Rossiter, "Bots on the Ground."
12. Rossiter, "Robotics, Autonomous Systems Warfare."
13. Amnesty International, *UN: Decisive Action Needed*.

14. Group of Governmental Experts, Report of the 2018 session, 5.
15. Ibid.
16. Cummings, *Artificial Intelligence and the Future of Warfare*, 4–6.
17. Rasmussen, "Skills, Rules, and Knowledge."
18. Cummings, *Artificial Intelligence and the Future of Warfare*, 5.
19. Ibid., 6.
20. Ibid., 7.
21. Smith, "The Miracle on the Hudson."
22. Cummings, *Artificial Intelligence and the Future of Warfare*, 6.
23. Ibid. 7.
24. Group of Governmental Experts of the High Contracting Parties to the Convention on Prohibitions or Restrictions on the Use of Certain Conventional Weapons Which May be Deemed to Be Excessively Injurious or to Have Indiscriminate Effects, Report of the 2018 Session, 4.
25. Ibid., 5.
26. Group of Governmental Experts of the High Contracting Parties to the Convention on Prohibitions or Restrictions on the Use of Certain Conventional Weapons Which May be Deemed to Be Excessively Injurious or to Have Indiscriminate Effects, Report of the 2018 Session, 5.
27. Sharkey, "Autonomous Weapon Systems," 76.
28. Human Rights Watch, "Losing Humanity."
29. See Sharkey, "Autonomous Weapon Systems" for a review of different interpretations of 'human dignity.'
30. Johnson and Axinn, "The Morality of Autonomous Robots."
31. Ibid.
32. Report of the Special Rapporteur, 14.
33. Alston and Shamsi, "A Killer Above the Law."
34. Report of the Special Rapporteur, 15.
35. Ibid.
36. See note 28 above.
37. Ulgen, "Human Dignity in an Age of Autonomous Weapons," 5.
38. Ibid., 15.
39. Walzer, *Just and Unjust Wars*, 136.
40. Ulgen, "Human Dignity in an Age of Autonomous Weapons," 19.
41. Ibid., 7.
42. See for example: Enemark, *Armed Drones and the Ethics of War*.
43. Enemark, *Armed Drones and the Ethics of War*, 60.
44. Ulgen, "Human Dignity in an Age of Autonomous Weapons," 19.
45. Strawser, "Moral Predators."
46. See for example: Ulgen, "Human Dignity in an Age of Autonomous Weapons"; Heyns, "Autonomous Weapons in Armed Conflict"; Human Rights Watch, "Losing Humanity."
47. Ibid.
48. Sharkey, "Autonomous Weapon Systems," 83.
49. Pop, "Autonomous Weapon Systems."
50. Ibid., 84.
51. Walzer, *Just and Unjust Wars*, 21.
52. Ibid., 41.
53. Group of Governmental Experts of the High Contracting Parties to the Convention on Prohibitions or Restrictions on the Use of Certain Conventional

Weapons Which May be Deemed to Be Excessively Injurious or to Have Indiscriminate Effects, Report of the 2018 Session.
54. Ibid., 4.
55. Ibid., 7.
56. Henckaerts and Doswald-Beck, *Customary International Humanitarian Law*, 4.
57. Ibid., 11.
58. *Protocol Additional to the Geneva Conventions* (I).
59. Melzer, *Interpretative Guidance*, 4.
60. Ibid.
61. Ibid., 45.
62. Schmitt, "Deconstructing Direct Participation in Hostilities"; Melzer, *Interpretative Guidance.*
63. Rossiter, "Participation in Warfare."
64. Bodnar and Pacho, "Targeted Killings (Drone Strikes)," 185.
65. Powell, *My American Journey*, 144.
66. See Yousaf, "U.S. Drone Campaign in Pakistan's Pashtun 'Tribal' Region."
67. Heller, "One Hell of a Killing Machine."
68. Benson, "Kill 'Em and Sort It Out Later," 18.

Acknowledgments

The author is extremely grateful to the two reviewers for their valuable comments. The author is also grateful to the team behind this special issue, especially the guest editor Ash Rossiter.

Disclosure statement

There are no conflict of interest in publishing this article. All opinions are the author's own.

Bibliography

Alston, Philip, and Hina Shamsi. 2010. "A Killer above the Law." *The Guardian*, February 2.

Amnesty International. 2018. UN: Decisive Action Needed to Ban Killer Robots - before It's Too Late. press release. August 27.

Benson, Kristina. "Kill 'Em and Sort It Out Later: Signature Drone Strikes and International Humanitarian Law." *Pacific McGeorge Global Business and Development Law Journal* 17, no. 1 (2014): 17–51.

Bodnar, Adam, and Irmina Pacho. "Targeted Killings (Drone Strikes) and the European Convention on Human Rights." *Polish Yearbook of International Law* 32 (2013): 189–208.

Campaign to Stop Killer Robots. 2018. "The Problem." https://www.stopkillerrobots.org/learn/#problem

Colin, Powell. *My American Journey*. New York: Random House, 1996.

Cummings, Mary L. *Artificial Intelligence and the Future of Warfare*. London: Chatham House, 2017.

Enemark, Christian. *Armed Drones and the Ethics of War: Military Virtue in a Post-Heroic Age*. Abingdon: Routledge, 2014.

"Group of Governmental Experts of the High Contracting Parties to the Convention on Prohibitions or Restrictions on the Use of Certain Conventional Weapons Which May Be Deemed to Be Excessively Injurious or to Have Indiscriminate Effects." 2018. *Report of the 2018 session of the Group of Governmental Experts on Emerging Technologies in the Area of Lethal Autonomous Weapons Systems*, October 23. Geneva. CCW/GGE.1/2018/3.

Haas, Benjamin. 2018. "'Killer Robots': AI Experts Call for Boycott over Lab at South Korea University." *The Guardian*, April 5. https://www.theguardian.com/technology/2018/apr/05/killer-robots-south-korea-university-boycott-artifical-intelligence-hanwha

Heller, Kevin Jon. "One Hell of a Killing Machine: Signature Strikes and International Law." *Journal of International Criminal Justice* 11, no. 1 (2013): 94–103. doi:10.1093/jicj/mqs093.

Henckaerts, Jean-Marie, and Louise Doswald-Beck. *Customary International Humanitarian Law: Volume 1, Rules*. Cambridge: Cambridge University Press, 2005.

Heyns, Cristof. "Autonomous Weapons in Armed Conflict and the Right to a Dignified Life: An African Perspective." *South African Journal on Human Rights* 22, no. 1 (2017): 46–71. doi:10.1080/02587203.2017.1303903.

Human Rights Watch. 2012. "Losing Humanity: The Case Against Killer Robots." https://www.hrw.org/report/2012/11/19/losing-humanity/case-against-killer-robots

International Committee of the Red Cross, Protocol Additional to the Geneva Conventions of 12 August 1949, and Relating to the Protection of Victims of International Armed Conflicts.

Johnson, Aaron M., and Sidney Axinn. "The Morality of Autonomous Robots." *Journal of Military Ethics* 12, no. 2 (2013): 129–141. doi:10.1080/15027570.2013.818399.

Kahn, Paul. "The Paradox of Riskless Warfare." *Philosophy and Public Policy Quarterly* 22, no. 3 (2002): 2–7. doi:10.13021/G8pppq.222002.380.

Leveringhaus, Alex. 2017. "Autonomous Weapons Mini-Series: Distance, Weapons Technology and Humanity in Armed Conflict." *Humanitarian Law & Policy*, October 6. ICRC.

Levin, Sam, and Julia Carrie Wong. 2018. "Self-Driving Uber Kills Arizona Woman in First Fatal Crash Involving Pedestrian." *The Guardian*, March 19.

"Meeting of the High Contracting Parties to the Convention on Prohibitions or Restrictions on the Use of Certain Conventional Weapons Which May Be Deemed to Be Excessively Injurious or to Have Indiscriminate Effects." 2017. *Final Report*, December 11. CCW/MSP/2017/8.

Melzer, Nils. *Interpretative Guidance on the Notion of Direct Participation in Hostilities under International Humanitarian Law*. Geneva: ICRC, 2009.

"Open Letter to Professor Sung-Chul Shin, President of KAIST from Some Leading AI Researchers in 30 Different Countries." March 2018. https://www.cse.unsw.edu.au/~tw/ciair//kaist.html

Pop, Ariadna. 2018. "Autonomous Weapon Systems: A Threat to Human Dignity?" *Humanitarian Law & Policy*, April 10. ICRC.

Rasmussen, Jens. "Skills, Rules, and Knowledge: Signals, Signs, and Symbols, and Other Distinctions in Human Performance Models." *IEEE Transactions on Systems, Man, and Cybernetics* 13, no. 3 (1983): 257–266. doi:10.1109/TSMC.1983.6313160.

"Report of the Special Rapporteur on Extrajudicial, Summary or Arbitrary Executions." 2014. *UN General Assembly*, August 6. A/96/265.

Rossiter, Ash. "Participation in Warfare: Are We Witnessing Paradigmatic Change." *Journal of Military and Strategic Studies* 19, no. 3 (2019): 103–123.

Rossiter, Ash. 2020. "Robotics, Autonomous Systems Warfare." *Small Wars & Insurgencies* 31, no. 3.

Rossiter, Ash. 2020. "Bots on the Ground: An Impending UGV Revolution in Military Affairs?" *Small Wars & Insurgencies* 31, no. 3.

Saxton, Adam. 2016. "(Un)dignified Killer Robots? the Problem with the Human Dignity Argument." *Lawfare*, March 20.

Schmitt, Michael N. "Deconstructing Direct Participation in Hostilities: The Constitutive Elements." *International Law and Politics* 42 (2010): 697–739.

Sharkey, Amanda. "Autonomous Weapon Systems, Killer Robots and Human Dignity." *Ethics and Information Technology* 21, no. 2 (2019): 75–87. doi:10.1007/s10676-018-9494-0.

Smith, Adam. 2006. "The Miracle on the Hudson: How It Happened." *The Telegraph*, November 22. https://www.telegraph.co.uk/films/sully/miracle-on-the-hudson-how-it-happened/

Strawser, Bradley J. "Moral Predators: The Duty to Employ Uninhabited Aerial Vehicles." *Journal of Military Ethics* 9, no. 4 (2010): 342–368. doi:10.1080/15027570.2010.536403.

Ulgen, Ozlem. "Human Dignity in an Age of Autonomous Weapons: Are We in Danger of Losing an 'Elementary Consideration of Humanity'." *European Society of international Law Conference Paper Series, 8, no. 9, Conference paper No. 15/2016*, Riga.

Walzer, Michael. *Just and Unjust Wars: A Moral Argument with Historical Illustrations*. 4th ed. New York: Basic Books, 2006.

Yousaf, Farooq. "U.S. Drone Campaign in Pakistan's Pashtun 'Tribal' Region: Beginning of the End under President Trump?" *Small Wars & Insurgencies* 31, no. 3 (2020).

Friend or frenemy? The role of trust in human-machine teaming and lethal autonomous weapons systems

Aiden Warren and Alek Hillas

ABSTRACT
This article explores the imprecise boundary between Lethal Autonomous Weapons Systems (LAWS) and Human-Machine Teaming – as a subset of Human-Machine Interaction – and the extent both are emerging as a point of concern (and option) in military and security policy debates. As the development of Human-Machine Teaming relates to artificial intelligence (AI) capabilities there also exists an area of concern pertaining to reliability and confidence, particularly in the heat of battle. Also known as Manned-Unmanned Teaming, Human-Machine Teaming attempts to engender trust and collaborative partnerships with robots and algorithms. Clearly the prospect of LAWS in recent times, or so-called 'killer robots,' has raised questions relating to the degree such devices can be trusted to select and engage targets without further human intervention. Aside from examining the 'trust factor,' the article also considers security threats posed by both state and non-state actors and the complicit yet inadvertent role multinational corporations play in such developments where civilian technology is modified for dual-purposes. The effectiveness of government regulation over AI, including whether AI can be 'nationalised' for national security reasons, will also be examined as part of AI non-proliferation.

Introduction

Human-machine teaming, or manned-unmanned teaming, is a subset of human-machine interaction. It is a process to build and then utilise a partnership between people and artificial intelligence (AI) so that each party utilises its own strengths to achieve a shared goal. Combining the capabilities of humans and computers is significantly about understanding context and processing large amounts of information. Benefits include enhanced decision-making and situational analysis. For example, as humans play a vital role navigating through uncertain warfighting contexts, machines will apply computational abilities to recommend actions in time-critical situations.

It is crucial to understand how military personnel will interact with the very robots they will be held responsible for in the context of war. The military and strategic context for such considerations relates to how autonomous systems development will the shape the capability of high-tech states, such as the U.S., to offset the conventional forces of adversaries. The successful adoption of machine autonomy will largely depend on whether states can develop policies to ensure that collaboration between robots and people achieves mission objectives at the operational level. To a large extent, the direction that military planners choose to take in forthcoming years regarding funding and planning for the research and development of machine autonomy, is dependent on how constructively people engage with semi-autonomous weapon systems, as the precursor to Lethal Autonomous Weapons Systems (LAWS). This article explores the imprecise boundaries between LAWS and semi-autonomous devices, which will be reliant on effective human-machine teaming for reliability and performance. In this context, trust is one of the most important factors for the success of human-machine teaming.

That said, it is evident that the existence of such technology will also generate new risks, making the global strategic landscape less predictable and more volatile than it is today. Management of these risks will vary across political systems, with the article considering whether democracies are better positioned, compared with autocratic regimes, to invest in human-machine teaming. Moreover, the proliferation and know-how of such technology secured by non-state actors could pose a new type of asymmetrical threat to national security. Therefore, one of the greatest challenges for policy-makers today is clearly the integration of lethal autonomous robotics into the military domain, while simultaneously ensuring that know-how is safe-guarded from being transferred to traditional adversaries and non-state actors.

For the main extent, states' military objectives are to not harm civilians, which could attract penalties and disrupt the peacebuilding process. However, unlike nation states, non-state actors would not have the desire or adequate resources to invest in making robots comply with International Humanitarian Law (IHL) (also known as the laws-of-war or laws-of-armed-conflict). Non-state actors could see LAWS as an option to embolden the asymmetrical advantages they intermittently hold against conventional forces. Due to the accessibility of data and the dual-use of robotic devices, the possible acquisition and use of LAWS by non-state actors would pose a new threat to national security. The use of 'killer robots' on civilians in places of gathering could terrorise a generation that is no longer as shocked by improvised devices, mass shootings and vehicle-based attacks. Moreover, without needing to find and train new recruits (robots could simply be replaced), non-state actors could refine their skills and experience, shortening the time between attacks. In the context of states being largely unable to regulate AI within their borders or transnationally, the article also examines the implications of the proliferation of lethal algorithms to non-state actors.

The political context for such considerations has thus far positioned itself with the United Nations (UN) Convention on Certain Conventional Weapons (CCW). The UN held the first Meeting of Experts on LAWS in 2014. It elevated the issue to the Group of Governmental Experts (GGE) on LAWS, first held in 2017. Civil society advocates led by the Campaign to Stop Killer Robots, '[a] coalition of 112 non-governmental organizations in 55 countries,' are advocating for states to negotiate a new CCW protocol to ban LAWS.[1] Some of the technical challenges and concerns relating to LAWS pertain to the degree to which robots can distinguish between combatants and non-combatants and whether robots could apply the proportionate force to achieve military objectives. Other issues relate to how a human would be able to determine and address causes of error in real time; and mete out punishment for misconduct. If the international community does not adopt hard or soft law to address the research, development, use and transfer of LAWS technologies, it is unlikely states would exercise self-restraint in this area of disarmament diplomacy through bilateral treaties compared to nuclear weapons. Unlike nuclear weapons, where only a handful of countries possess stockpiles and the necessary materiel is difficult for others to acquire covertly, there would be fewer barriers to states, and even non-state actors, in gaining a foothold into LAWS technologies.

The article begins with an examination of the military incentives to develop human-machine teaming, followed by a section on the political incentives across different systems of government. The final component of the article considers available options for addressing proliferation risks of dual-use technologies to adversaries, including state and non-state actors.

Role of trust in the human-robot team

Humans use trust as a substitute for total reliability. Although trust can feel like an unquantifiable 'gut feeling,' it is measurable and is an essential element for teamwork. Trust can exist within teams, between teams, and in technology. It also encompasses factors such as honesty, understanding, respect, quality of interaction, reliability, communication, ability, performance and expectancy.[2] If a team member or operational process is no longer trustworthy, time and other resources (such as money) need to be spent to evaluate and implement an alternative method to achieve the same objective. For example, consider a Military Working Dog team, consisting of a human and the canine as a semi-autonomous weapon. If the handler is conducting an investigation into contraband, they will rely on their canine to establish probable cause in the search and 'should be aware of the dog's successful training exercises as well as the dog's actual record of success in similar search situations.'[3] This relates to trust. While we do not fully understand the cognitive processes of animals – or even our own cerebral functions – it is possible to develop trust through substitutes such as knowledge of someone or something's training and demonstrated abilities.

How does trust apply to robots? Humans do not trust machines in the same way that we trust each other (or even some intelligent animals). According to Kristin Schaefer, the four types of human-to-robot trust include:

- the propensity to trust, which is unique to the individual human;
- cognition-based trust, which changes over time as humans learn more about how the robot works, how to interact with the robot, and related expectations;
- affect-based trust, which measures how people make attributions about the robot based on human emotive factors including comfort, satisfaction, confidence and attitude; and
- initial trustworthiness, which is based on the robot's reputation, physical appearance, 'personality,' 'intelligence,' level of automation and perceived function.[4]

One state's key document that considers human oversight is available from the United Kingdom Ministry of Defence. In May 2018, the Development, Concepts and Doctrine Centre published *Joint Concept Note 1/18 Human-Machine Teaming*.[5] In a submission to the GGE in August 2018, the UK Government refers to this Joint Concept Note as 'the UK Ministry of Defence's thinking on the evolution of how humans will work with machines.'[6] Among its findings, the UK identifies 'what [human] actors trust their machines to do' as something that will restrain 'the increasing capabilities of robotic and AI systems.' It categorises four 'fundamental factors' that will determine how much human actors are prepared to trust remote and automated systems: mechanical understanding, predictability, familiarity and context.[7]

Under this model of human-machine teaming, people will need to have technical expertise as well as emotional intelligence. For example, operators will need to not only 'understand how a system works' on an intellectual level, but also recognise their own emotional responses, such as confidence. Moreover, the actor's ability to forecast the likely errors of a machine will also depend on their (and their machines') previous training and experiences in the field – something that seems possible for the military to refine over time. That said, on a political level, these systems must also be trusted by 'observers' to work effectively.[8] (These observers are, presumably, members of issues-motivated groups or civil society who have an interest in armed conflict and advocate at the domestic and international level.) While the most vocal observers today oppose or express serious concerns with the prospect of lethal robotics, the UK has inferred that such 'norms of trust and public appetites' could shift over time.[9] That is, if highly automated weapons comply with IHL and there is public acceptance of this, states could even be pressured to increase the autonomy levels of their robots. The claim, made in *JCN 1/18*, of the possibility that humans could perceive robots differently in the future compared to today, remains a proposition that will only be able to be tested upon further advances in technology. To reach that stage would require demonstrating how human-machine teaming could work at the operational level, to which the next section turns.

Computers already have the capacity to use algorithms to generate more data about an environment, at speeds faster than humans can process the same information. Conversely, humans continue to understand context better than machines and can use this to analyse situations with reason, not computational logic. These two styles seem incompatible. Yet, they are complementary. Bob Work, in 2015 as the U.S. Deputy Secretary of Defense under President Obama, spoke about the role of human-machine collaboration in the context of the Obama administration's Third Offset Strategy. Work noted the surprising victory of Steven Cramton and Zackary Stephen, two amateur chess players who used their personal computers to defeat supercomputers and grandmasters in a chess competition in 2005, as an example of 'allowing a machine to help humans make better decisions faster' and using 'machines to help our decision-makers make better decisions.'[10] Computer-assisted decision-making in combat is already available in the form of the F-35, which collects and integrates information for presentation to the pilot. As described by Work, 'The F-35 is not a fighter plane. It is a flying sensor computer that sucks in an enormous amount of data, correlates it, analyzes it, and displays to the pilot on his helmet.'[11] The U.S. is not the only nation interested in this space. At the GGE held in March 2019, the Netherlands, although aligning itself with the EU Statement,[12] indicated that is had explored in a national capacity 'the possible benefits of human-machine collaboration,' including 'the optimal utilization of technological benefits, such as precision, speed and reliability without sacrificing the robustness and flexibility of human intelligence.'[13] In this regard, human-machine teaming appears well-positioned to benefit from AI without becoming beholden to it.

So, how can or should this work? In June 2015, the Office of the Chief Scientist of the United States Air Force released a vision document on the role of human-machine teaming in airpower, titled, *Autonomous Horizons: System*

Autonomy in the Air Force – A Path to the Future, Volume 1: Human-Autonomy Teaming. It notes, 'airmen [will] need to develop informed trust – an accurate assessment of when and how much autonomy should be employed, and when to intervene.'[14] This includes three factors: Systems factors, Individual factors, and Situational factors. As an integrated assessment, this would combine and replace the separate decisions that would have been made in the past pertaining to Information ('how much to trust information inputs'), Others ('how must to trust other teammates') and Automation ('how much to trust a particular automated tool').[15] The difficulty lies in ensuring that the human actor does not over-trust *or* under-trust autonomous systems. As a means to address dichotomy, the document suggested adopting guidelines to make it easier to determine trustworthiness levels. This would encompass systems designed to assist with decision-making by reasoning in a similar way to humans. In practice this means coding simple, logical rules rather than a deep-learning neural network. Other recommendations included: joint training, avoiding any human-like features in the system to deter against over-confidence, and providing contextual overviews and visualisations to share situational awareness.[16] The second volume of *Autonomous Horizons*, published in 2019, provides an update and notes the Air Force will concentrate on four areas or 'tenants of trust.' These are cognitive congruence and transparency, situation awareness, human-systems integration, and human-system teaming and training. These tenants must be considered 'from the beginning of the design process' to maximise trust in machine-assisted decision-making.[17]

The amount of trust that a person has in a robot will depend on whether the commander considers that the machine is enhancing or limiting their Course of Action (COA) decisions. COAs are a request for authorisation to undertake an action. The number of options, presented to the human for decision, also informs the nature of interaction. As the German delegation to the March 2019 GGE said:

> The unique qualities of human judgement can't be totally replaced by the capacities of machines, which have a high capacity for analyzing large sets of mathematical data but which can't be trusted to take the kind of value based decisions which military practitioners are required to take under international law.[18]

In exercising judgement, then, 'human control can only be assured as long as the respective accountable human has sufficient knowledge of the machine, of the operating environment and of the likely interaction between the two.'[19] The important consideration is for the person to *actually* make a decision and not just give authorisation to any proposed action.

Trust is a special factor for determining the number of robots under human supervision in a swarm. A program by the U.S. has examined the number of viable COA decisions available with a swarm size of greater than 100 drones.[20] Such a program would need to be able to factor in the baseline skills of any person who is making the decisions or tailor the complexity of the COAs to a specific individual. Currently, technical challenges to interacting with a machine include unintuitive multisensory interfaces, unreliable natural language processing, ineffective gesture control, insufficient trust in machine autonomy, and immature decision support tools. If these issues are resolved, it may become possible for the military to test out teaming with highly autonomous robots. The objective is 'safe and effective human-machine systems successfully operating in complex, dynamic and contested environments.' An estimated timeframe posited by the Office of Naval Research is around 2026.[21] Of course, this raises the question of how the U.S. envisions the role of trust in human-machine teaming within that period.

Armin Krishnan was one of the first to propose that robots could be restricted to a 'kill box' – geographical coordinates which have been pre-selected by humans in advance, where there are little to no civilians – in defensive circumstances.[22] Similarly, Ronald Arkin points to measures that could reduce the likelihood of friendly fire,[23] while Zachary Morris, writing in *Military Review*, has proposed that methods used to confirm that civilians are not present in the designated geographical kill box could be 'similar to clearing areas for artillery fire.'[24] Accordingly, a staged approach that takes into account the limitations of robotics would be a 'responsible crawl-walk-run methodology [that] allows for experimental improvement and increased experience while adequately protecting innocent lives and clarifying responsibility,' according to Morris.[25] Nevertheless, the creation of a kill box would not absolve responsibilities under national and international law. In a publication on Joint Fire Support, the U.S. Department of Defense (DoD) has defined the purpose of a kill box as 'a measure, not a mission' and cautions that 'Fires executed in a kill box should comply with ROE [Rules of Engagement] and law-of-war targeting constraints; designation of a kill box is not authorization to fire indiscriminately into the area.'[26] This calls into question whether the use of autonomous weapons could be a realistic option for kill box commanders, particularly in relation to how confident the commander would be in making a legal order.

At the GGE in August 2018, the U.S. delegation highlighted its research into the challenges associated with opaque trust and accountability. The Explainable Artificial Intelligence (XAI) program by the Defense Advanced Research Projects Agency (DARPA) seeks to address this.[27] According to publicly available information, the XAI program focuses on both the human and the machine. It aims:

> to create a suite of machine learning techniques that: produce more explainable models, while maintaining a high level of learning performance (prediction

> accuracy); and enable human users to understand, appropriately trust, and effectively manage the emerging generation of artificially intelligent partners.[28]

This deals with how a computer can describe its methods and conclusion to a human, including "the psychology of explanation,'[29] and would help to explain concepts in a way that should be familiar to the end-user, who is not an expert in computers.[30] David Gunning, a program manager for XAI with a background in both computer science and cognitive psychology (who had previously also managed the program that produced Siri), presented at the Institute of International and European Affairs about the XAI program in December 2018. Gunning noted how DARPA's two challenge problems would have two military applications:

- First, to allow defence analysts to feel more confident when they agree to a computer's recommendation – cognizant that if something goes wrong, the analyst is generally responsible for accepting the decision (rather than the programmer for coding an algorithm that wasn't explainable to the end-user).
- Second, military operators would be able to ask a computer to explain why it made decisions in the field and evaluate the impact of its decisions.[31]

In expressing concerns pertaining to the linkage between automation and complex security scenarios, the International Committee of the Red Cross (ICRC) has stated the aim of the XAI program may ultimately be 'to apply machine learning to automatic target recognition and identification.' This 'promises a capability advantage for decision-making in data-rich conflict environments [...] despite the risks of unpredictability, which may conflict with military commanders' propensity for command and control.'[32] In this analysis, the concept of human-machine teaming may have strategic benefits but could also lead to some tactical missteps. At the GGE held in March 2019, the ICRC also spoke in general terms about how machine learning has 'unpredictability by design' because 'machine-learning algorithms [...] are "black boxes" and their functioning cannot be explained.' This problem is worsened by testing that cannot verify performance but can 'introduce bias, whether by design or through bias in the data used to "train" (develop) the algorithms.'[33] The analysis here suggests that there is a trade-off between whether the unpredictability and bias of machines is made up for by the promised capability advantage for decision-making.

Moreover, Emma Schuster argues that there are trade-offs associated with 'value-alignment' – 'a key area of AI research which looks at finding ways to ensure that AI systems can recognise and align their behaviour with human values ... [which] forces a trade-off between the ability of the system to optimise its output.' In this analysis, the right balance or 'trade-off' between human oversight and machine performance is struck through value-alignment.[34] The characteristic of value-alignment is similar with 'the underlying logic of IHL generally.' That is, IHL constrains some possible uses of force, just as the decision to not utilise the mathematically incomprehensible logic of AI deep learning may prevent the computer program reaching its potential. Here, the central issue pertains to where the balance lies between trusting a machine and accepting liability for unknown errors, versus a better understanding of its (comparatively limited) algorithmic pathways. In March 2019, the GGE discussions noted that: 'A related characteristic of importance [to *predictability*] is *explainability*, whether a system allows an operator to understand why it performs in a particular way. *Reliability*, whether the operator can trust the system to perform consistently as intended by the commander, is also crucial.'[35] Optimistic of striking such a balance, Poland has suggested that 'predictability [...] could be ensured by establishing human parenting to supervise and control the learning process with code of conduct deriving from IHL, in which case the predictability [of proper functioning of a weapon system, particularly the target, time and location of an attack] might be bigger than in humans.'[36] Given the above-mentioned issues, there is a strong military incentive to continue investing in the human element of machine autonomy in the form of human-machine teaming. The next section compares the political incentives across systems of government.

Incentives to invest in human-machine teaming across systems of government

Of course, there is a high likelihood that the purpose and utility of LAWS would differ across democratic and autocratic political systems. As most states have not called for a ban, it is evident that LAWS could have different perceived benefits across various systems of government. Arguably, for the purposes of this article, the GGE provides an up-to-date account on state thinking about human-machine teaming. (Interestingly, there was an even spread among the countries that had called for a ban on LAWS by November 2018.[37] According to the 2019 Freedom in the World Survey, nine were ranked 'free', eight 'partly free', and nine 'not free'.)[38]

Democracies, more than any other system of government, are motivated by a 'push factor' to reduce their military casualties. As a consequence of the requirement for 'large-coalition leaders [to] keep a broad swath of the people happy,' democratic leaders are prepared to spend more on each soldier, providing better training and equipment, 'so they can achieve the same military output while putting few soldiers at risk.'[39] Nevertheless, casualty rate increases over the duration of any conflict can lead to public fatigue with protracted operations. Perhaps as a prominent example, during the 2016 U.S. federal election the Republican Party's campaign may have benefitted from 'a significant and meaningful relationship between a community's rate of military sacrifice and its support for Trump.'[40] For example in early 2019 Trump said that the withdrawal from Syria would fulfil a campaign promise 'to bring our troops back home from endless wars. [...] They've been going on for 19 years in the area.'[41] While there is some continuity with the Obama administration's wind-back of military engagements, under Trump a greater emphasis has been placed on the capabilities that the military requires to succeed.[42] Trump's first federal budget proposal asked Congress to '[provide] for one of the largest increases in defense spending [...] offset and paid for by finding greater savings and efficiencies across the Federal Government.'[43] The 'America First Budget' noted that, 'Unlike spending increases for war, which mostly consume resources in combat, the increases in the President's Budget primarily would invest in a stronger military ... and would ensure the Armed Forces have the training, equipment, and infrastructure they need.'[44] While this shift has contributed to strained relations with some allied nations, it arguably responds to populist concerns about reducing military casualties.

Conversely, a 'pull factor' regarding machines is that people are biased in believing that technology – including military technology – is superior to human capabilities.[45] The existence of uninhabited systems, like drones, is making it increasingly harder for politicians to justify combat deaths. As part of the Third Offset Strategy under President Obama, the U.S. was seeking to develop 'collaborative human-machine operations' which are 'about making the human operate better' by forming a new type of team that can 'offset' conventional forces. This included humans continuing to 'be the ones who make decisions on lethal force.'[46] The potential casualty-reduction envisaged was captured in a quote attributed to Deputy Secretary Work: 'I'm telling you right now, 10 years from now if the first person through a breach isn't a **** robot, shame on us.'[47] The policy approach was in accordance with Department of Defense Directive 3000.09 (set to expire in 2022), which mandates that a human must make the decision about using lethal force, with the exception of autonomous cyber weapons and returning force to another autonomous system.[48] As such, while it may be inaccurate to suggest that robots will wholly replace American soldiers on the battlefield, it is plausible to say that robots will play a larger role, particularly in reducing combat casualties.

Under Trump, there has been little to no mention of the Third Offset Strategy by name due to its close association with the Obama administration. Further, as an attempt to 'move away from a predominantly technology-first approach to defence innovation,' Fiott notes that the Department of Defense has spent:

> much more time to thinking about operational and tactical needs first and which technologies could be adapted for strategic purposes later ... [T]here is some way to go before the DoD and the U.S. military fully integrate the human–machine relationship into their operational doctrines and institutional cultures.[49]

Comparing the changed priorities between the two presidents, Fiott concludes that, although the challenges which the Third Offset Strategy sought to address remain prevalent, 'the seeming shift from technology-centric thinking within the DoD to more operationally minded considerations may suggest that the U.S. has reached a more mature stage in thinking about defence.'[50] Due to the superpower status of the U.S. (which has not called for a ban on LAWS), the effect of democracy on autonomous weapons policy, and specifically, the role of trust in human-machine teaming, is one of the most important areas for consideration as a possible precursor to LAWS.

In the context of autocratic systems, robots could serve a very different purpose. In this regard, such autonomous devices could further embolden despots to rule without compromise and without the fear of needing to negotiate with others (even the moderates in their own party who may seek to grant concessions in exchange for public legitimacy). If this is the case, LAWS could be a cost-saving device for dictators. Autocrats currently need to buy the loyalty of an army to carry out unpleasant or extremely violent tasks – such as the suppression of peaceful protesters, where robots could be more reliable than humans. Funds saved from not needing to 'buy' the loyalty of the army and essential political backers could be redirected to increase the private wealth of the ruler and build a financial buffer to draw on when the regime's survival is truly on the line.

In international armed conflicts, autocrats would not have as much incentive to develop human-machine teaming to cut casualties. As the authors of *The Dictator's Handbook* contend: 'In autocracies, foot soldiers are not politically important. Autocrats do not waste resources protecting them.'[51] According to selectorate theory, autocratic systems of government are run by political leaders who need to distribute private rewards to their essential backers to stay in power. In wartime, these leaders must choose between continuing to fund their backers or diverting money towards the war effort. For example, an analysis of the Six-Day War found President Nasser of Egypt 'would likely have lost the loyal support of lots of his key backers if he took their 150,000 USD a head and spent it on the war instead of on them. He actually would have increased his chance of being overthrown in a military coup by making an all-out effort to win the war at the expense of his cronies.' In contrast, the relevant Israelis would have had to sacrifice less than 59 USD.[52] Thus, there is a clear distinction between political systems for rationalising the expenses to invest in human-machine teaming, depending on the political cost of military casualties.

Most concerning is that, in contrast to LAWS used in warfare, where the principle of distinction between combatants and non-combatants is a critical component of IHL, robots tasked by an autocrat to use hostile force against protestors domestically would not be required to have a level of sophistication. Clearly, it would be counterproductive for such actors to invest their resources in developing improvements to the technology to meet International Human Rights Law (IHRL). In addition, non-state actors could rely on 'conventional techniques that are known to be crude but adequate,' also a concern for additive manufacturing (3-D printing).[53] In attempting to devise their own LAWS, terrorists could seek to convert commercially available uninhabited devices into relatively unsophisticated killer robots that inflict mass casualties. The means to do so may be available from the apparent openness of the AI community or for sale on the dark web. As precursory examples, the relatively low-tech disruption of London's Gatwick and Heathrow airports in December 2018 and January 2019 respectively, are two of the latest instances where authorities had failed to keep up to date with technologies available, whether commercially or otherwise. The relative simplicity of LAWS that do not comply with IHL and IHRL means that these will be able to be developed before the use of sophisticated LAWS in armed conflicts, suggesting the acquisition of LAWS by autocrats and non-state actors is a more pressing global security concern. The role of the tech industry in addressing such threats posed by non-state actors is explored in greater detail in the next section.

Trusting the tech industry to address the risks of non-proliferation?

Having considered the accessibility and uses of lethal robotics above, the pertinent question remains as to whether such devices could be kept out of reach of non-state actors or adversaries, particularly as dual-use technologies become more readily available. Russian President Vladimir Putin has said that whoever monopolises AI 'will become the ruler of the world.'[54] Putin's statement raises the question: Can AI be nationalised? The answer may vary across the major powers. Clearly, Russia's leader considers that AI should be nationalised. Washington has arguably been unsuccessful at coercing tech giants to respond to its national security threats. As an example, following the December 2015 San Bernardino shooting, an extremist-motivated attack that resulted in the most fatalities in the U.S. since 9/11,[55] Apple opposed a court order requiring it to assist an FBI search of the iPhone of one of the terrorists. In February 2016, Apple CEO Tim Cook published 'A Message to Our Customers,' outlining the company's objections 'for the government to force us to build a backdoor into our products.'[56] Acknowledging the balance between protecting civil liberties and protecting against security threats, President Barack Obama criticised Apple's objections as an 'absolutist view' that risked 'fetishizing our phones above every other value. And that can't be the right answer.'[57] (Once the government successfully accessed the data stored – without Apple – it asked for the order requiring Apple's assistance to be vacated.) To explore recent activities in this area, we examine Google, as a well-known multinational, and consider whether advanced, open economies like the U.S. can have oversight or control of possible dual-purpose technologies developed inside their borders by private sector industries. This has implications for whether non-state actors could have access to sensitive technology or the infrastructure required to adapt code into something compatible with LAWS.

In the digital age, the proliferation of any algorithm is possible in the form of the creation or transmittal of algorithms that may be adapted for use by robotic devices. The computing power required for training an AI, through machine learning, is disproportionately large compared to what is needed to run the program afterward. According to the UN Institute for Disarmament Research (UNIDIR), 'This means that while the data and computing power needed to create the most powerful new AI algorithms may be limited to a smaller set of actors, a larger number of groups and individuals may often be able to use already-trained algorithms.'[58] In addition, the United Nations Office for Disarmament Affairs (UNODA) notes in terms of security that, '[t]he pervasive nature of digital technologies may exacerbate ... risks, including by non-State actors.'[59] This suggests that the international community needs to respond to the prospect of technological proliferation, including risks to global security posed by modified robotics.

To understand what some of the numbers could involve, consider the power required to train and then run the AlphaZero algorithm, developed by Google DeepMind. In 2017, AlphaZero used self-play reinforcement learning to train 'itself' to play chess and defeat the world champion in four hours. Some of the hardware required was 5,000 first-generation Tensor Processing Units (TPUs) – Google's custom-designed, application-specific integrated circuit. This was complimented by 64 second-generation TPUs for training the algorithm's neural networks.[60] After 'it' completed this learning, the AlphaZero algorithm only needed 4 TPUs, comparatively little computing power, to play chess during a competition. For context, the first-generation TPU was 15–30 times more powerful, as well as 30–80 times more energy efficient (as measured by performance per watt), than the benchmark Graphics Processing Units and Central Processing Units that were available in 2015.[61] As an alternative to volunteer computing, which requires large public appeal for scientific research,[62] anyone with the right amount of cash can train an algorithm on a platform, such as Google Cloud,[63] provided they meet the Terms of Service. The price for the public to 'rent' a single second-generation TPU in December 2019 began at 1.35 USD per hour.[64] This raises the question of whether the increasing availability of cloud-based infrastructure could remove a barrier for anyone – including non-state actors – to do intensive training for algorithms.

The acquisition of algorithms that have already been trained may be available publicly or through the dark web. Indeed, as noted by Munich-based Conscious Coders, hardware and instructions for building autonomous systems is readily accessible.[65] Of course, the threat posed by non-state actors is determined by their capability and intent. As Chertoff elucidates,

> [While s]cholars and officials have expressed concern over online access to drone blueprints and machine learning algorithms, which non-state actors could use to build autonomous weapons [...] to effectively innovate these technologies internally, non-state actors still require the tacit knowledge [such as experience or intuition from working with the technology].[66]

As with any other successful insurgency, non-state actors would need to be capable across a variety of areas – ranging from combat, to finance, recruitment, medical, transport and logistics. However, what is evident is that such proliferation concerns, whether through the creation or the transmittal of hardware and software, will only intensify in the future.

Indeed, proliferation concerns were discussed at the GGE of August 2018, where states adopted ten 'Possible Guiding Principles.'[67] The GGE emphasized the need for international law to play a more decisive role:

> When developing or acquiring new weapons systems based on emerging technologies in the area of LAWS, physical security, appropriate non-physical safeguards (including cyber-security against hacking or data spoofing), the risk of acquisition by terrorist groups and the risk of proliferation should be considered.[68]

The GGE agreed its ongoing work should be driven by IHL, the UN Charter and pertinent ethical viewpoints. According to the Chair of the 2019 GGE, in future discussions, 'Other types of risks should be considered, as appropriate, including the risk of unintended engagements, loss of control of the system to unauthorized parties, risks of proliferation, acquisition by terrorist groups, and potential effects on the security environment.'[69] António Guterres, the UN Secretary-General, has attempted to give strategic direction to this issue. In an address to the General Assembly in September 2018, Guterres said he was 'encouraged by the ten possible guiding principles.' Guterres also set his expectations for member states: 'Let's call it as it is. The prospect of machines with the discretion and power to take human life is morally repugnant.'[70] At the beginning of the March 2019 session of the GGE, Guterres reiterated his call about 'politically unacceptable, morally repugnant' machines and advocated they 'should be prohibited by international law.'[71] Arguably, the international community has not heeded its Secretary-General's call, which indicates the complexity of the issue. It is necessary, then, to examine events taking place in other fora – namely, the tech industry.

Tech workers, scientists and members of civil society have made headlines in their attempt to emphasise the sense of urgency associated with emerging technologies. In May 2017, the inaugural session of the GGE, scheduled to take place in August 2017, was cancelled. The cancellation was due to unpaid contributions to UN disarmament conferences by member states. (Remarkably, Brazil had not paid for seven years.)[72] Noting the cancellation, on the day that the GGE had been due to commence leaders of companies who considered themselves engaged in 'building the technologies in Artificial Intelligence and Robotics that may be repurposed to develop autonomous weapons' urged swift diplomatic resolution in an open letter published through the Future of Life Institute (FOLI).[73] In July 2018, FOLI published a 'LAWS Pledge,' in which over 160 AI-related organisations, including Google DeepMind, undertook to 'neither participate in nor support the development, manufacture, trade, or use of lethal autonomous weapons.'[74] This built on an open letter about autonomous weapons published by FOLI in July 2015, which had warned that '[i]t will only be a matter of time until they appear on the black market and in the hands of terrorists, dictators wishing to better control their populace, [and] warlords wishing to perpetrate ethnic cleansing.'[75]

Such open letters have created a space for lower-level employee activists to emerge.[76] For instance, the Arms Control Association 2018 Arms Control Person of the Year award recognised 'a group of 4,000 anonymous Google employees opposing the company's work on a Pentagon project using artificial intelligence.'[77] In April 2018, the Google employees wrote anonymously to their CEO, Sundar Pichai, about how their company's work could be used 'by U.S. Government drones to detect vehicles and other objects, track their motions, and provide results to the Department of Defense.' The employees objected to Google's alleged involvement 'in the business of war.' They asked to cancel the contract and called for 'a clear policy stating that neither Google nor its contractors will ever build warfare technology. Warning of reputational damage to Google's 'brand and its ability to compete for talent [...] Google will join the ranks of companies like Palantir, Raytheon, and General Dynamics," which the signatories alleged were companies that had lost public trust.[78] Amr Gaber, one of the signatories of the April 2018 letter (and a volunteer with the Tech Workers Coalition), spoke at a side event of the GGE in August 2018. Graber noted that "technology gets more opaque through layers of automation, the cloud, and the artificial intelligence and machine learning technologies." As such, technology is being developed by people who don't know the "true nature of the project" they are working on.[79] In June 2018, Pichai outlined a new set of AI principles or "concrete standards that will actively govern [Google's] research and product development and will impact our business decisions."

The AI Principles would prevent Google from "developing AI for use in weapons."[80] In response, Bob Work, who in 2017 announced Project Maven as using AI to search for objects of interest located in photos and videos,[81] defended the potential benefit to save American and allied lives and juxtaposed the Google employees' apparent silence over the recent opening of the Google AI China Center.[82]

While the above incident demonstrates that the pressure to act on autonomous weapons is dynamic – beginning with top-down pressure from industry leaders who established an ethical compass on the issue through FOLI, then shifting to pacifist employees who made their case for change from the bottom-up – the effectiveness of Google's AI Principles was called into question after several months. The latter development increased focus over the question of whether Google (as one example of a large tech company) is required to act loyally to any one country in respect to the development of potentially sensitive technology, as well as how states can respond to allegations that AI is being nationalised by an adversary. In August 2018, Google employees signed an open letter about the alleged lack of ethical responsibility in their industry, specifically, the reported development of 'Dragonfly,' which the open letter signatories described as 'an effort to provide Search and personalised mobile news to China, in compliance with Chinese government censorship and surveillance requirements.' The signatories brought to attention: 'That the decision to build Dragonfly was made in secret, and progressed even with the AI Principles in place makes clear that the Principles alone are not enough. We urgently need more transparency, a seat at the table, and a commitment to clear and open processes: Google employees need to know what we're building.'[83] If such an app were designed to comply with Chinese Mainland laws, including Beijing's Cybersecurity Law, Google users' locations and private data would be 'much more vulnerable to government access,' according to Amnesty International.[84]

For additional context in relation to the opaqueness of potential dual-use technologies, the U.S. National Defense Strategy 'recognizes the challenges presented by rapid technological advancements in dual-use areas, and China's blurring of the lines between civil and military goals.'[85] In its 2019 Annual Report to Congress on China, the DoD suggests that AI – including 'unmanned systems, human-machine teaming, swarming technology, and lethal autonomous weapons' – is being nationalised by Beijing. Aided by '[t]he country's effort to build national corporate champions that achieve rapid market dominance across a range of frontier technologies [which] directly complements the PLA's [People's Liberation Army's] modernization,'[86] the national civil-military integration (CMI) strategy elevated by President Xi in 2015 acts as:

> a national strategy focused on aligning civil and defense technology development to achieve greater efficiency, innovation, and growth [...] harnessing

> emerging dual-use technologies such as AI, machine learning, big data, and unmanned systems to facilitate what PLA writings refer to as 'intelligentized' warfare, or using multiple data streams and information flows to enable PLA operations.[87]

This assessment of Beijing's focus on CMI raises the question of whether the U.S., in response, is in a position to 'nationalise' dual-use technologies developed inside its borders.

Soon after Google employees signed the 'Dragonfly' letter, Vice-President Mike Pence called on Google to 'immediately end development of the "Dragonfly" app that will strengthen Communist Party censorship and compromise the privacy of Chinese customers.'[88] Pichai testified about Dragonfly before the House Judiciary Committee in December 2018.[89] Following a Senate Armed Services Committee Hearing in March 2019 where U.S. defence leaders said that Google already provided 'direct benefit' to the Chinese military,[90] President Trump tweeted that 'Google is helping China and their military.'[91] Yet, after Trump met with Pichai in March 2019, one of their positions changed (though it is not exactly clear whose), with Trump tweeting Pichai had 'stated strongly that he is totally committed to the U.S. Military, not the Chinese Military.'[92] The Dragonfly incident is a recent and applicable example of how the U.S. could respond to private companies seen to be developing dual-purpose technologies with benefits to foreign powers. In addition, when Maven and Dragonfly are viewed together as a case study, employee motivations are contrasted with corporate goals. Neither align with the state's strategy, suggesting there will be continued difficulty for the U.S. to nationalise its burgeoning tech sector. While a military superpower would not be challenged directly by a superpower computer, some services made available to the public by tech companies, if unregulated, could make it easier for nefarious actors to threaten national security with emerging technologies.

Pertinent to the question of whether AI can be nationalised are domestic legislation and the global governance rules that enable transnational corporations (TNCs) to structure across multiple jurisdictions and operate across borders, often synonymous with the free-flow of finance and human capital. In 2016, political movements across the Atlantic pushed back against some elements of globalisation. In the Brexit Referendum, the UK voted to 'leave the European Union,'[93] after the Leave campaign pledged to 'Take Control.'[94] In the U.S. Federal Election, Donald Trump promised to 'drain the swamp.'[95] During his inauguration, he said, 'we are not merely transferring power from one Administration to another, or from one party to another – but we are transferring power from Washington, D.C. and giving it back to you, the American People.'[96] As a result of such populist based assertions (and perhaps to safeguard against protectionism), companies are required to be seen not to act against the national interest. Nevertheless, with further trade liberalisation still underway in the form of the World Trade Organisation's 'Trade in Services Agreement,' the U.S. and European Union's 'Transatlantic Trade and Investment Partnership' and the China-led 'Free Trade Area of the Asia Pacific,' which are all currently being negotiated,[97] as well as the 'Regional Comprehensive Economic Partnership' agreed in late 2019 which will be signed following a technical legal review, TNCs can expect to continue to operate across relatively open borders. The above examples of dual-use technology suggest companies and governments alike are unequipped to deal with the national security elements posed by the broadening globalisation of the international political economy.

Conclusion

Clearly the prospect of LAWS in recent times, or so-called 'killer robots,' has raised questions relating to the degree such devices can be trusted to select and engage targets without further human intervention. Aside from examining the 'trust factor,' this article also considered security threats posed by both state and non-state actors and the complicit yet inadvertent role multinational corporations play in such developments where civilian technology could be modified for dual-purposes. The effectiveness of government regulation over AI, including whether AI can be 'nationalised' for national security reasons, was also examined as part of AI non-proliferation.

The article first examined publicly known military research projects that have attempted to improve human trust in computers. As indicated, trust can be defined in operational contexts, including the degree to which we can actually 'place trust in a robot's recommended course of action.' Robots and computers are, relative to humans, bad at explaining why one decision is preferable to another. This is a major problem that will need to be overcome to reap the perceived benefits of AI-assisted decision-making, if this is at all attainable. Given the investment of appropriate resources, human trust in machines may be forced to grow steadily over time. As such, the article assessed the extent to which robots would be able to wholly replace human judgement, or remain as devices that enhance and compliment. The outcome of course will have direct implications for how robots will be deployed and utilized in both military and commercial spaces.

The article also considered the degree to which the tech industry can be trusted in preventing the proliferation of harmful technologies to non-state actors and whether the state, as a response, could 'nationalise' companies that produce AI-related technologies. As indicated, UN members are aware of the risks posed by terrorists and how commercial autonomous technologies could be repurposed and weaponised. Yet there is little consensus about a multilateral response pertaining to this threat. Domestic responses would be varied. For countries like China that pursue civilian-military integration in which the private and public sectors are blurred, it appears that AI is already being nationalised. In contrast, democracies seem vulnerable insofar as it is more difficult in coercing transnational companies toward promoting one government's agenda. That said, Washington's response, as evident in Congressional hearings and varying fora with the Executive Branch, indicates that America has a system perhaps more conductive to a workable, mutually beneficial relationship with such entities.

While the handful of states calling for a pre-emptive ban on LAWS includes a wide array of governments, most of the international community is leaving the use of LAWS open as a possibility. Clearly the purpose of acquiring LAWS would vary across political systems. Democracies, motivated to reduce military casualties, would need to invest significant resources to achieve the high levels of human-machine teaming required to make use of lethal robotics. In contrast, authoritarian rulers are more reliant on the support of key backers rather than the public. In non-democratic systems, LAWS could be a cost-saving tool for dictators who want to suppress protesters when segments of the police or army are resistant to using lethal force. Arguably, the multilateral forums held on LAWS to date have focused more on warfighting scenarios than crack-downs on popular movements. Yet, for regulation to be effective on the global level and to address proliferation to non-state actors, regulation must address why LAWS are being pursued under each political system.

While the rapid development of LAWS appears highly desirable to proponents of robotic systems, their expansion has engendered significant apprehension among diplomats, human rights campaigners, technology and arms control advocates, and others who argue that deploying fully autonomous weapons in conflicts would severely diminish human oversight during operations. Such overtures could plausibly lead to the contravention of the laws of war, and could undermine the barriers that mitigate escalation from conventional to nuclear war. Indeed, the emergence of new technologies in the form of LAWS has spurred questions pertaining to their legality, morality, and broader wisdom in their overall deployment. This debate has gained traction as the United States, Russia, and several other states have amplified their expansion of such weapons systems, with each signifying what they must do so as to thwart their rivals from gaining an advantage in these new methods of warfare.[98] In considering some of these aforementioned challenges and concerns, this article focused on the role of trust and applications of human-machine teaming, the purposes of LAWS in various systems of government, and overall the degree to which these may present a challenge to an already complex and ambiguous security environment.

Notes

1. Campaign to Stop Killer Robots, "Intervention by the Campaign," 1.
2. Ashley & Stanton cited in Stanton, "An Introduction to Trust," 1–2.
3. Bernstein (ed.), *The Military Commander*, 199.

4. Schaefer, "The Perception and Measurement," 33–38.
5. UK MoD, *Joint Concept Note 1/18*, 48.
6. UK, "Human Machine Touchpoints," 2.
7. UK MoD, *Joint Concept Note 1/18*, 49.
8. Ibid.
9. Ibid.
10. Work, "The Third Offset Strategy." See also: Scharre, "The Opportunity and Challenge," 17.
11. Work, "The Third Offset Strategy."
12. EU, *Statement*.
13. Netherlands, *Statement*, 3.
14. USAF Office of the Chief Scientist, *Autonomous Horizons*, 8.
15. Ibid 9.
16. Ibid 19–20.
17. Zacharias, *Autonomous Horizons: The Way Forward*, 84.
18. Germany, *Statement*, 2.
19. Ibid.
20. Chung, "OFFensive Swarm-Enabled Tactics," 5.
21. Tangney, "Human Systems Roadmap Review," 26–27.
22. Krishnan, "Automating War," 188–189. See also: Krishnan, *Killer Robots*, 162; Lin, Bekey & Abney, *Autonomous Military Robots*, 77; Arkin, *Governing Lethal Behaviour*, 171–176.
23. Arkin, *Governing Lethal Behaviour*, 155, 169, 171, 173.
24. Morris, "A Four-Phase Approach," 4.
25. Ibid 8.
26. U.S. DoD, *Joint Publication 3–09*, A-9. (Bold in original.)
27. U.S, *Human-Machine Interaction*, 5.
28. Turek, "Explainable Artificial Intelligence."
29. Ibid.
30. Gunning, "XAI," 7:04–8:44.
31. Ibid 12:35–15:11.
32. ICRC, "Ethics and autonomous weapon systems," 15–16.
33. ICRC, "Statement," 4.
34. Schuster, "When Something Has To Give," 52.
35. Gjorgjinski, "Chair's non-paper," 6.
36. Poland, *Statement*.
37. Campaign to Stop Killer Robots, "Country Views," 1–2.
38. Freedom House, "Freedom in the World Countries." Note: Palestine and the Holy See were not in the survey.
39. Bueno de Mesquita & Smith, *The Dictator's Handbook*, 229–230.
40. According to analysis based on an online representative sample of 2,148 U.S. citizens, conducted in November 2015 and published in September 2016. Kriner & Shen, "Battlefield Casualties and Ballot Box Defeat," 2.
41. Trump, "Remarks."
42. Fiott, "America First," 47.
43. Trump, "America First," 1–2.
44. Office of Management and Budget, *America First*, 15–16.
45. Schneider & Macdonald, *U.S. Public Support*, 15.
46. Work, "Remarks."

47. Work cited by Sadowski, "Enabling MUM-T," 2. See also, Work, "The Third Offset Strategy."
48. Schroeder, "Policies," 38, 52–53.
49. Fiott, "America First," 46–47.
50. Ibid 48.
51. Bueno de Mesquita & Smith, *The Dictator's Handbook*, 230.
52. Ibid 11–12, 198–199, 213, 219, 222, 232.
53. Brockmann & Kelley, *The Challenge of Emerging Technologies*, 36.
54. RT, "Whoever leads in AI."
55. U.S. Extremist Crime Database cited in U.S. Government Accountability Office, *Countering Violent Extremism*, 34.
56. Cook, *A Message*.
57. Obama, "Remarks."
58. UNIDIR, *The Weaponization of Increasingly Autonomous Technologies*, 4.
59. UNODA, *Securing Our Common Future*, 51.
60. Silver et al, "Mastering Chess and Shogi," 4.
61. Sato, Young & Patterson, "AI & Machine Learning." For a technical explanation, see Jouppi et al, "In-Datacenter Performance Analysis."
62. BOINC, "Volunteer computing."
63. Google Cloud, "Platform Terms of Service."
64. Google Cloud, "TPU pricing." Note: 'In order to connect to a TPU, you must provision a virtual machine (VM), which is billed separately. For details on pricing for VM instances, see Compute Engine pricing.'
65. Pohler et al, "A Technological Perspective," 13.
66. Chertoff, *Perils of Lethal Autonomous Weapons Systems Proliferation*, 9.
67. CCW, *Tweet: 10 Possible Guiding Principles*.
68. Ibid.
69. Gjorgjinski, "Chair's non-paper," 5.
70. Guterres, *Address*.
71. CCW, *Tweet: Message by @UN Secretary-General*.
72. UN, *2017 Group of Governmental Experts*. See also: UN, *Status of contributions*, 3; Campaign to Stop Killer Robots, *Diplomatic Efforts Falter*.
73. FOLI, "An Open Letter."
74. FOLI, "About the LAWS Pledge. See also, FOLI, Lethal Autonomous Weapons Pledge."
75. FOLI, "Autonomous Weapons."
76. Gaber, "Transcript of UN Remarks."
77. Arms Control Association, "Group of 4,000 Anonymous Google Employees."
78. 'Googlers', "Letter to Google."
79. Gaber, "Transcript of UN Remarks," 2.
80. Pichai, "AI at Google."
81. U.S. DoD, "Project Maven."
82. Peniston, "Google's Withdrawal from Pentagon."
83. Menegus, "Here's the Letter."
84. Amnesty International, "Open letter."
85. U.S. DoD, *Assessment on U.S. Defense Implications*, 24. See also: U.S. DoD, *Summary of the 2018 National Defense Strategy*, 15.
86. U.S. DoD, *Annual Report to Congress*, 101–102.
87. Ibid 21.
88. Pence, "Remarks."

89. C-Span, "Google Data Collection," 2:08:25–2:12:30.
90. Hawley, "Following Google CEO Meeting with President Trump."
91. Trump, "Tweet: Google is helping China."
92. Trump, "Tweet: Just met with @SundarPichai."
93. UK National Archives, *European Union Referendum Act 2015*. See also: BBC News, *EU Referendum Results*.
94. Vote Leave, *Campaign Resources*.
95. Hughes, "Trump calls to "drain the swamp"."
96. Trump, "The Inaugural Address."
97. Yueh, *The Great Economists*, 283–286.
98. Klare, "Autonomous Weapons Systems."

Disclosure Statement

No conflicts of interest were declared for the publication of this article.

Funding

The authors received no financial support for the publication of this article.

Bibliography

Amnesty International. "Open Letter: Google Must Not Capitulate on Human Rights to Gain Access to China." Accessed August 28, 2018. http://web.archive.org/web/20200104004545/https://www.amnesty.org/en/latest/news/2018/08/open-letter-to-google-on-reported-plans-to-launch-a-censored-search-engine-in-china/

Arkin, Ronald. *Governing Lethal Behaviour in Autonomous Robots*. Boca Raton, FL: CRC Press, 2009.

Arms Control Association. *Group of 4,000 Anonymous Google Employees Urging Company Not to Be 'In the Business of War' Voted 2018 Arms Control Persons of the Year*. Arms Control Association, 10 January 2019. http://web.archive.org/web/20191214161545/https://www.armscontrol.org/pressroom/2018-acpoy-winner

BBC News. "EU Referendum Results." http://web.archive.org/web/20200103060403/https://www.bbc.com/news/politics/eu_referendum/results

Berkeley Open Infrastructure for Network Computing. "Volunteer Computing." University of California, 2018. http://web.archive.org/web/20200103214757/https://boinc.berkeley.edu/trac/wiki/VolunteerComputing

Bernstein, Graham H., ed. *The Military Commander and the Law*. 13th ed. Maxwell Air Force Base, AL: The Judge Advocate General's School, 2016. http://web.archive.org/web/20200102224513/https://www.holloman.af.mil/Portals/101/documents/JA%20Documents/References/MCL%202016%20web.pdf?ver=2016-11-28-145828-903

Brockmann, Kolja, and Robert Kelley. *The Challenge of Emerging Technologies to Non-Proliferation Efforts: Controlling Additive Manufacturing and Intangible Transfers of Technology*. Stockholm: Stockholm International Peace Research Institute, 2018. http://web.archive.org/web/20200103042102/https://www.sipri.org/sites/default/files/2018-04/sipri1804_3d_printing_brockmann.pdf

Bueno de Mesquita, Bruce, and Alastair Smith. *The Dictator's Handbook: Why Bad Behaviour Is Almost Always Good Politics*. New York: PublicAffairs, 2011.

Campaign to Stop Killer Robots. "Diplomatic Efforts Falter." May 30, 2017. http://web.archive.org/web/20200103223943/https://www.stopkillerrobots.org/2017/05/diplomatsfalter/

Campaign to Stop Killer Robots. "Country Views on Killer Robots." Accessed November 22, 2018. web.archive.org/web/20191224114734/https://www.stopkillerrobots.org/wp-content/uploads/2018/11/KRC_CountryViews22Nov2018.pdf

Campaign to Stop Killer Robots. "Intervention by the Campaign to Stop Killer Robots." Presented for informal consultation to the Group of Governmental Experts on Emerging Technologies in the Area of Lethal Autonomous Weapons Systems, Geneva, Accessed June 28, 2019. http://web.archive.org/web/20200102223734/https://www.stopkillerrobots.org/wp-content/uploads/2019/06/KRC_StatementCCW_28June2019_Delivered.pdf

CCW @CCWUNODA. "Tweet: 10 Possible Guiding Principles 4 Emerging Technologies in the Area of Lethal Autonomous Weapons Systems Affirmed by the Group of Governmental Experts on Certain Conventional Weapons @ the Conclusion of Its Meeting on 31 August 2018 @ungeneva." Twitter, Accessed September 1, 2018. http://web.archive.org/web/20200103221754/https:/twitter.com/CCW_UNODA/status/1035938908482232321?s=20

CCW @CCWUNODA. "Tweet: Message by @UN Secretary-General @antonioguterres (Delivered by @UNOG_DG) to the Meeting of the #CCWUN Group of Governmental Experts on Emerging Technologies in the Area of Lethal Autonomous Weapons Systems (#LAWS) Meeting @UNgeneva This Week." Twitter, Accessed March 25, 2019. http://web.archive.org/web/20200103222809/https://twitter.com/CCW_UNODA/status/1110187627427311618

Chertoff, Philip. *Perils of Lethal Autonomous Weapons Systems Proliferation: Preventing Non-State Acquisition*. Geneva Centre for Security Policy, Accessed Oct. 2018, 2. http://web.archive.org/web/20200103221102/https://dam.gcsp.ch/files/2y10RR5E5mmEpZE4rnkLPZwUleGsxaWXTH3aoibziMaV0JJrWCxFyxXGS

Chung, Timothy H. "OFFensive Swarm-Enabled Tactics (OFFSET)." Presented at the DARPA Tactical Technology Office Proposers Day, Arlington, VA, Accessed January 30, 2017. http://web.archive.org/web/20200102235700/https://www.darpa.mil/attachments/OFFSET_ProposersDay.pdf

Cook, Tim. *A Message to Our Customers*. Apple, Accessed February 16, 2016. http://web.archive.org/web/20200103044437/https://www.apple.com/customer-letter/

C-SPAN. "Google Data Collection." Accessed December 11, 2018. https://www.c-span.org/video/?455607-1/google-ceo-sundar-pichai-testifies-data-privacy-bias-concerns&start=7813

European Union. "Statement." Presented at the Group of Governmental Experts on Emerging Technologies in the Area of Lethal Autonomous Weapons Systems, Geneva, March 25–29, 2019. http://web.archive.org/web/20200102233039/https://www.unog.ch/80256EDD006B8954/%28httpAssets%29/1F4322BC2BCC4EF3C12583CB003B39F9/$file/ALIGNED+-+LAWS+GGE+EU+statement+Human+Element.pdf

Fiott, Daniel. "America First, Third Offset Second?" *The RUSI Journal* 163, no. 4 (2018): 40–48. doi:10.1080/03071847.2018.1529893.

Freedom House. "Freedom in the World Countries." http://web.archive.org/web/20200103031105/https://freedomhouse.org/report/countries-world-freedom-2019?order=field_fiw_aggregate_score&sort=asc

Future of Life Institute. "About the Lethal Autonomous Weapons Systems (LAWS) Pledge." https://futureoflife.org/laws-pledge/

Future of Life Institute. "An Open Letter to the United Nations Convention on Certain Conventional Weapons." http://web.archive.org/web/20200103224209/https://futureoflife.org/autonomous-weapons-open-letter-2017

Future of Life Institute. "Autonomous Weapons: An Open Letter from AI & Robotics Researchers." http://web.archive.org/web/20200103225210/https://futureoflife.org/open-letter-autonomous-weapons

Future of Life Institute. "Lethal Autonomous Weapons Pledge." http://web.archive.org/web/20200103224917/https://futureoflife.org/lethal-autonomous-weapons-pledge/

Gaber, Amr. "Transcript of UN Remarks." Presented at a side event of the Group of Governmental Experts on Emerging Technologies in the Area of Lethal Autonomous Weapons Systems, Geneva, Accessed August 28, 2018. http://web.archive.org/web/20200103232943/https://www.stopkillerrobots.org/wp-content/uploads/2018/11/Amr-Gaber-UN-remarks-2018-28-8-1.pdf

Germany. "Statement." Presented at the Group of Governmental Experts on Emerging Technologies in the Area of Lethal Autonomous Weapons Systems, Geneva, March 25–29, 2019. http://web.archive.org/web/20200102235434/https://www.unog.ch/80256EDD006B8954/%28httpAssets%29/2B8E772610C0F552C12583CB003A4192/$file/20190326+Statement3+Germany+GGE+LAWS.pdf

Gjorgjinski, Ljupčo Jivan. "Chair's Non-paper of Draft Conclusions." Presented for an informal meeting of the CCW on May 15, 2019, following on from the Group of Governmental Experts on Emerging Technologies in the Area of Lethal Autonomous Weapons Systems, Geneva, March 25–29, 2019. http://web.archive.org/web/20200103014636/https://www.unog.ch/80256EDD006B8954/(httpAssets)/D8C04EC71F502A77C12583F400476619/$file/Draft+Conclusions+(ϕ).pdf

Google Cloud. "Platform Terms of Service." Accessed November 21, 2019. http://web.archive.org/web/20200103215345/https://cloud.google.com/terms/

Google Cloud. "TPU Pricing." Accessed December 5, 2019. http://web.archive.org/web/20200103220251/https://cloud.google.com/tpu/pricing

Googlers. "Letter to Google." Letter Linked by Shane Scott and Daisuke Wakabayashi, "'The Business of War': Google Employees Protest Work for the Pentagon," *New York Times*, Accessed April 4, 2018. http://web.archive.org/web/20200104000355/https://static01.nyt.com/files/2018/technology/googleletter.pdf

Gunning, David. "XAI: Can Artificial Intelligence Be Made Intelligible?" Presented at the Institute of International and European Affairs, Dublin, Accessed December 5, 2018. http://web.archive.org/web/20200103011046/https://www.iiea.com/past-events/xai-can-artificial-intelligence-be-made-intelligible/ (YouTube: https://youtu.be/0np9N0C5ukc)

Guterres, António. "Address to the General Assembly." Presented at the United Nations, New York, Accessed September 25, 2018. http://web.archive.org/web/20200103222500/https://www.un.org/sg/en/content/sg/speeches/2018-09-25/address-73rd-general-assembly

Hawley, Josh D. "Following Google CEO Meeting with President Trump, Senator Hawley Calls on Google to Publicly Address Its Partnership with China." Accessed March 28, 2019. http://web.archive.org/web/20190329105236/https://www.hawley.senate.gov/following-google-ceo-meeting-president-trump-senator-hawley-calls-google-publicly-address-its

Hughes, Trevor. "Trump Calls to 'Drain the Swamp' of Washington." *USA Today*, Accessed October 18, 2016. http://web.archive.org/web/20200103061109/https://www.usatoday.com/story/news/politics/elections/2016/2016/10/18/donald-trump-rally-colorado-springs-ethics-lobbying-limitations/92377656/

International Committee of the Red Cross. "Ethics and Autonomous Weapon Systems: An Ethical Basis for Human Control?" Presented at the Group of Governmental Experts on Emerging Technologies in the Area of Lethal Autonomous Weapons Systems, Geneva, April 9–13, 2018. http://web.archive.org/web/20200103013445/https://www.unog.ch/80256EDD006B8954/%28httpAssets%29/42010361723DC854C1258264005C3A7D/$file/CCW_GGE.1_2018_WP.5+ICRC+final.pdf

International Committee of the Red Cross. "Statement." Presented at the Group of Governmental Experts on Emerging Technologies in the Area of Lethal Autonomous Weapons Systems, Geneva, March 25-29, 2019. http://web.archive.org/web/20200103013801/https://www.unog.ch/80256EDD006B8954/%28httpAssets%29/A8474F2C2ED6CD48C12583CC00300940/$file/CCW+GGE+LAWS+ICRC+statement+agenda+item+5b+26+03+2019.pdf

Jouppi, Norman P., Cliff Young, Nishant Patil, David Patterson, Gaurav Agrawal, Raminder Bajwa, Sarah Bates. "In-Datacenter Performance Analysis of a Tensor Processing Unit™." Paper presented at the 44th International Symposium on Computer Architecture, Toronto, June 24–26, 2017. http://web.archive.org/web/20200103065658/https://arxiv.org/ftp/arxiv/papers/1704/1704.04760.pdf

Klare, Michael T. "Autonomous Weapons Systems and the Laws of War." *Arms Control Today*, Accessed March 2019. http://web.archive.org/web/20200104024018/https://www.armscontrol.org/act/2019-03/features/autonomous-weapons-systems-laws-war

Kriner, Douglas, and Francis Shen. "Battlefield Casualties and Ballot Box Defeat: Did the Bush-Obama Wars Cost Clinton the White House?" *Available* at SSRN 2017. 10.2139/ssrn.2989040

Krishnan, Armin. "Automating War: The Need for Regulation." *Contemporary Security Policy* 30, no. 1 (2009): 172–193. doi:10.1080/13523260902760397

Krishnan, Armin. *Killer Robots: Legality and Ethicality of Autonomous Weapons*. Surrey, England/ Burlington, VT: Ashgate, 2009.

Lin, Patrick, George Bekey, and Keith Abney. *Autonomous Military Robots: Risks, Ethics and Design*. San Luis Obispo, CA: California Polytechnic State University, 2008.

Menegus, Bryan. "Here's the Letter 1400 Google Workers Sent Leadership in Protest of Censored Search Engine for China." *Gizmodo*, August 17, 2018. http://web.archive.

org/web/20200104003819/https://www.gizmodo.com.au/2018/08/heres-the-letter-1400-google-workers-sent-leadership-in-protest-of-censored-search-engine-for-china/

Morris, Zachary L. "A Four-Phase Approach to Developing Ethically Permissible Autonomous Weapons." *Military Review*, May 2018. http://web.archive.org/web/20190728055630/https://www.armyupress.army.mil/Portals/7/Army-Press-Online-Journal/documents/Morris-v2.pdf

Netherlands. "Statement." Presented at the Group of Governmental Experts on Emerging Technologies in the Area of Lethal Autonomous Weapons Systems, Geneva, March 25–29, 2019. http://web.archive.org/web/20200102233602/https://www.unog.ch/80256EDD006B8954/%28httpAssets%29/164DD121FDC25A0BC12583CB003A99C2/$file/5b+NL+Statement+Human+Element-final.pdf

Obama, Barack H. "Remarks." Presented at South by Southwest Interactive, Austin, TX, March 11, 2016. http://web.archive.org/web/20200103045137/https://obamawhitehouse.archives.gov/the-press-office/2016/03/14/remarks-president-south-southwest-interactive

Office of Management and Budget. *America First: A Budget Blueprint to Make America Great Again*. Washington D.C.: Office of Management and Budget, 2017. http://web.archive.org/web/20200103032653/https://www.whitehouse.gov/wp-content/uploads/2017/11/2018_blueprint.pdf

Pence, Mike R. "Remarks by Vice President Pence on the Administration's Policy toward China." Presented at the Hudson Institute, Washington D.C., October 4, 2018. http://web.archive.org/web/20200104004756/https://www.whitehouse.gov/briefings-statements/remarks-vice-president-pence-administrations-policy-toward-china/

Peniston, Bradley. "Google's Withdrawal from Pentagon AI Project Risks US Lives, Says Work." *Defense One*, June 26, 2018. http://web.archive.org/web/20200105010406/https://www.defenseone.com/technology/2018/06/googles-withdrawal-pentagon-ai-project-risks-us-lives-bob-work/149280/

Pichai, Sundar. "AI at Google: Our Principles." June 7, 2018. http://web.archive.org/web/20200104001549/https://www.blog.google/technology/ai/ai-principles/

Pohler, Lukas, Valentin Schrader, Alexander Ladwein, and Florian von Keller. "A Technological Perspective on Misuse of Available AI." Consciouscoders, August, 2018. http://web.archive.org/web/20190728062003/https://www.consciouscoders.io/wp-content/uploads/2018/08/A_Technological-Perspective_on_Misuse_of_Available_AI-1.pdf

Poland. "Statement." Presented at the Group of Governmental Experts on Emerging Technologies in the Area of Lethal Autonomous Weapons Systems, Geneva, March 25–29, 2019. web.archive.org/web/20200105022832/https://www.unog.ch/80256EDD006B8954/%28httpAssets%29/5828C60CDCED2EF8C12583CC004C546F/$file/2.+GGE_LAWS_March+2019_PL+Statement_5%28c%29+Human+element_26.03.2019.pdf

RT (Russia Today). "'Whoever Leads in AI Will Rule the World': Putin to Russian Children on Knowledge Day." September 1, 2017. http://web.archive.org/web/20200103042903/https://www.rt.com/news/401731-ai-rule-world-putin/

Sadowski, Robert W. "Enabling MUM-T within Army Formations." Presented at Army Robotics Community of Practice, n.p, January 30, 2016. http://web.archive.org/web/20190728054345/https://www.darpa.mil/attachments/DARPA_MUMT_Updates.pdf

Sato, Kaz, Cliff Young, and David Patterson. "AI & Machine Learning: An In-depth Look at Google's First Tensor Processing Unit (TPU)." *Google Cloud*, May 13, 2017. http://web.archive.org/web/20200103070049/https://cloud.google.com/blog/products/gcp/an-in-depth-look-at-googles-first-tensor-processing-unit-tpu

Schaefer, Kristin E. "The Perception and Measurement of Human-Robot Trust." PhD diss., University of Central Florida, 2013. http://web.archive.org/web/20190728054759/http://etd.fcla.edu/CF/CFE0004931/Schaefer_Kristin_E_201308_PhD.pdf

Scharre, Paul D. "The Opportunity and Challenge of Autonomous Systems." In *Autonomous Systems: Issues for Defence Policymakers*, edited by Andrew P. Williams and Paul D. Scharre, 3–26. Norfolk, VA: NATO Communications and Information Agency, 2015. http://web.archive.org/web/20190728055234/http://www.act.nato.int/images/stories/media/capdev/capdev_02.pdf

Schneider, Jacquelyn, and Julia Macdonald. *U.S. Public Support for Drone Strikes: When Do Americans Prefer Unmanned over Manned Platforms?* Washington D.C.: Center for a New American Security, 2016. http://web.archive.org/web/20200103033659/https://s3.amazonaws.com/files.cnas.org/documents/CNAS-Report-DronesandPublicSupport-Final2.pdf

Schroeder, Ted W. "Policies on the Employment of Lethal Autonomous Weapon Systems in Future Conflicts." Masters thesis, American University, 2016. http://web.archive.org/web/20200103041843/https://www.secnav.navy.mil/innovation/Documents/2017/07/LAWS_Essay.pdf

Schuster, Emma. "When Something Has to Give: The Intersection of Artificial Intelligence, Military Decision-Making and International Humanitarian Law" Masters thesis, Sweden: Lund University, 2018. http://web.archive.org/web/20200103014120/http://lup.lub.lu.se/luur/download?func=downloadFile&recordOId=8955234&fileOId=8955235

Silver, David, Thomas Hubert, Julian Schrittwieser, Ioannis Antonoglou, Matthew Lai, Arthur Guez, Marc Lanctot et al. "Mastering Chess and Shogi by Self-Play with a General Reinforcement Learning Algorithm." *arXiv preprint* (2017). http://web.archive.org/web/20200103064821/https://arxiv.org/pdf/1712.01815.pdf

Stanton, Neville A. "An Introduction to Trust in Military Teams." In *Trust in Military Teams: Human Factors in Defence*, edited by Neville A. Stanton, 1–9. Boca Raton, FL: CRC Press, 2011.

Tangney, John. "Human Systems Roadmap Review." Presented at the Human Systems Conference, Springfield VA, February 9–10, 2016. http://web.archive.org/web/20200103004947/https://ndiastorage.blob.core.usgovcloudapi.net/ndia/2016/Human/S&TPanel.pdf

Trump, Donald J. "America First: Beginning a New Chapter of American Greatness." In *America First: A Budget Blueprint to Make America Great Again*, edited by Executive Office of the President of the United States, 1–2. Washington D.C: Office of Management and Budget, 2017. http://web.archive.org/web/20200103032653/https://www.whitehouse.gov/wp-content/uploads/2017/11/2018_blueprint.pdf

Trump, Donald J. "Remarks." Presented in Cabinet Meeting, Washington, D.C., January 2, 2019, http://web.archive.org/web/20200103032206/https://www.whitehouse.gov/briefings-statements/remarks-president-trump-cabinet-meeting-12/

Trump, Donald J. "The Inaugural Address." Presented at United States Capitol, Washington, D.C., January 20, 2017. http://web.archive.org/web/20200103061941/https://www.whitehouse.gov/briefings-statements/the-inaugural-address/

Trump, Donald J. "Tweet: Google Is Helping China and Their Military, but Not the U.S." Twitter, March 17, 2019. http://web.archive.org/web/20200104012043/https:/twitter.com/realdonaldtrump/status/1107025689486737410

Trump, Donald J. "Tweet: Just Met with @sundarpichai, President of @google, Who Is Obviously Doing Quite Well." Twitter, March 28, 2019. http://web.archive.org/web/20200104022751/https:/twitter.com/realDonaldTrump/status/1110989594521026561

Turek, Matt. "Explainable Artificial Intelligence (XAI)." Defense Advanced Research Projects Agency. http://web.archive.org/web/20190728055815/https://www.darpa.mil/program/explainable-artificial-intelligence

U.S. (United States). "Human-Machine Interaction in the Development, Deployment and Use of Emerging Technologies in the Area of Lethal Autonomous Weapons Systems." Presented at the Group of Governmental Experts on Emerging Technologies in the Area of Lethal Autonomous Weapons Systems, Geneva, August 27–31, 2018. http://web.archive.org/web/20200103010057/https://www.unog.ch/80256EDD006B8954/%28httpAssets%29/D1A2BA4B7B71D29FC12582F6004386EF/$file/2018_GGE+LAWS_August_Working+Paper_US.pdf

U.S. Air Force Office of the Chief Scientist. *Autonomous Horizons: System Autonomy in the Air Force – A Path to the Future*. Washington, D.C.: United States Air Force, March 2019. http://web.archive.org/web/20200102233854/https://www.af.mil/Portals/1/documents/SECAF/AutonomousHorizons.pdf?timestamp=1435068339702

U.S. Department of Defense. *Annual Report to Congress: Military and Security Developments Involving the People's Republic of China,* 2019. Washington, D.C.: Office of the Secretary of Defense. http://web.archive.org/web/20200103043825/https://media.defense.gov/2019/May/02/2002127082/-1/-1/1/2019_CHINA_MILITARY_POWER_REPORT.pdf

U.S. Department of Defense. *Assessment on U.S. Defense Implications of China's Expanding Global Access*. 2018. https://media.defense.gov/2019/Jan/14/2002079292/-1/-1/1/EXPANDING-GLOBAL-ACCESS-REPORT-FINAL.PDF

U.S. Department of Defense. *Joint Publication 3-09 – Joint Fire Support*. Washington, D.C.: U.S. Department of Defense, 2019. http://web.archive.org/web/20200103005822/https://www.jcs.mil/Portals/36/Documents/Doctrine/pubs/jp3_09.pdf

U.S. Department of Defense. "Project Maven to Deploy Computer Algorithms to War Zone by Year's End." July 21, 2017. http://web.archive.org/web/20200104000146/https://www.defense.gov/Explore/News/Article/Article/1254719/project-maven-to-deploy-computer-algorithms-to-war-zone-by-years-end/

U.S. Department of Defense. *Summary of the 2018 National Defense Strategy of the United States of America: Sharpening the American Military's Competitive Edge*. Washington, D.C.: U.S. Department of Defense, 2018. http://web.archive.org/web/20200103050453/https://dod.defense.gov/Portals/1/Documents/pubs/2018-National-Defense-Strategy-Summary.pdf.

U.S. Government Accountability Office, *Countering Violent Extremism: Actions Needed to Define Strategy and Assess Progress of Federal Efforts*. 2017. http://web.archive.org/web/20200103044156/https://www.gao.gov/assets/690/683984.pdf

United Kingdom. "Human Machine Touchpoints: The United Kingdom's Perspective on Human Control over Weapon Development and Targeting Cycles." Presented at the Group of Governmental Experts on Emerging Technologies in the Area of Lethal Autonomous Weapons Systems, Geneva, August 27–31, 2018. http://web.archive.org/web/20200102230939/https://www.unog.ch/80256EDD006B8954/%

28httpAssets%29/050CF806D90934F5C12582E5002EB800/%24file/2018_GGE+LAWS_August_Working+Paper_UK.pdf

United Kingdom Ministry of Defence. *Joint Concept Note 1/18 Human-Machine Teaming*. Shrivenham, Swindon, Wiltshire: United Kingdom Ministry of Defence, 2018. http://web.archive.org/web/20200102225357/https://assets.publishing.service.gov.uk/government/uploads/system/uploads/attachment_data/file/709359/20180517-concepts_uk_human_machine_teaming_jcn_1_18.pdf

United Kingdom National Archives. "European Union Referendum Act 2015." https://www.legislation.gov.uk/ukpga/2015/36

United Nations. "2017 Group of Governmental Experts on Lethal Autonomous Weapons Systems (LAWS)." https://web.archive.org/web/20170730042556/http://unog.ch/80256EE600585943/(httpPages)/F027DAA4966EB9C7C12580CD0039D7B5?OpenDocument

United Nations. "Status of Contributions of BWC, CCW, CCM, OTW as at 30 April 2017." http://web.archive.org/web/20200103223755/https://www.unog.ch/80256EDD006B8954/%28httpAssets%29/8CF0896B98B17936C1258114005D7324/$file/Status+of+Contrib+BWC+CCW+CCM+OTW_30+Apr+2017.pdf

United Nations Institute for Disarmament Research. *The Weaponization of Increasingly Autonomous Technologies: Artificial Intelligence; a Primer for CCW Delegates*. Geneva: United Nations Institute for Disarmament Research, 2018. http://web.archive.org/web/20200103063452/https://unidir.org/files/publications/pdfs/the-weaponization-of-increasingly-autonomous-technologies-artificial-intelligence-en-700.pdf

United Nations Office for Disarmament Affairs. *Securing Our Common Future: An Agenda for Disarmament*. New York: United Nations Office for Disarmament Affairs, 2018. http://web.archive.org/web/20200103064021/https://s3.amazonaws.com/unoda-web/wp-content/uploads/2018/06/sg-disarmament-agenda-pubs-page.pdf

Vote Leave. "Campaign Resources." http://web.archive.org/web/20200103060852/http://www.voteleavetakecontrol.org/campaign_resources.html

Work, Bob. "Remarks by Deputy Secretary Work on Third Offset Strategy." Presented to the European Policy Centre, Brussels, Belgium, April 28, 2016. http://web.archive.org/web/20200103034423/https://www.defense.gov/Newsroom/Speeches/Speech/Article/753482/remarks-by-d%20eputy-secretary-work-on-third-offset-strategy/

Work, Bob. "The Third Offset Strategy." Presented at the Reagan Defense Forum, Simi Valley, CA, November 7, 2015. http://web.archive.org/web/20200102232101/https://www.defense.gov/Newsroom/Speeches/Speech/Article/628246/reagan-defense-forum-the-third-offset-strategy/

Yueh, Linda. *The Great Economists: How Their Ideas Can Help Us Today*. United Kingdom: Viking, 2018.

Zacharias, Greg. *Autonomous Horizons: The Way Forward*. Maxwell Air Force Base, AL: Air University Press, 2019. http://web.archive.org/web/20200102234454/https://media.defense.gov/2019/Mar/18/2002102574/-1/-1/0/b_0155_zacharias_autonomous_horizons.pdf

Bots on the ground: an impending UGV revolution in military affairs?

Ash Rossiter

ABSTRACT

There has been speculation for some time that unmanned ground vehicles (UGVs) are poised to revolutionize military land operations. These expectations have amplified with recent developments, not least the reported deployment and testing of Russian weaponized UGVs in Syria. Yet when it comes to the operational use of mobile ground-based robots – armed or otherwise – the recent history of the technology can be described as one of promise so far unfulfilled. By tracing past and present efforts to develop and field UGVs – and the enduring challenges that lie therein – this article attempts to gauge the likely impact of such systems in future conflict, as well as their effect on international security more broadly. The article concludes that although UGVs will almost certainly become a major – if not indispensable – feature of future military land operations, they will, similar to other promising militarily relevant technologies before them, continue to produce unrealistic expectations about their impending revolutionary effect.

Introduction

That robotic and autonomous systems are changing the character and perhaps even the nature of war has become an all but pervasive idea in security studies literature[1] and within policy circles.[2] Given the ever-increasing use of drones in many states' military operations, much of the attention by scholars, think tank experts and pundits on robots in warfare has unsurprisingly focused on the use and proliferation of unmanned aerial systems (UASs).[3] But military experts and scholars have long been interested in the potential of unmanned ground vehicles (UGVs) to improve performance in a range of military tasks and to reduce risk to personnel.[4]

Predictions about the impending substitution of UGVs for manned systems – often in the form of warnings about the impending presence of 'killer robots' – have been with us for some time.[5] But attention on UGVs – and, concomitantly, debates surrounding their use – has intensified in recent years, not least because of fast-paced developments in artificial intelligence (AI) and the interrelated concern that nations might soon deploy ground-based lethal *autonomous* weapon systems (LAWS) in combat.[6] Notwithstanding strong ethical and legal objections to LAWS in some countries, most commentators assume that combat UGVs with increasingly levels of autonomy are set to become a significant means by which states produce violence on the battlefield.[7] Seemingly, this view is shared by senior officials in the world's leading capitals. In 2013, for example, the chief of the Russian military, Valery Gerasimov, posited that 'in the near future it is possible a fully robotized unit will be created, capable of independently conducting military operations.'[8] Breathing life into these prophesies, a number of governments are aggressively pursuing weaponized UGV programs.

Though expectations of widespread UGV use in warfare have been around for some time, there are, beyond those performing some niche tasks, relatively few actual systems fielded by military organizations today. Involvement in lengthy – and, in the case of Afghanistan, enduring – conflicts certainly generated incentives among participating states to transfer combat risk from humans to robotic systems.[9] And although this partly drove the rapid expansion of drone usage in twentieth-first century conflict,[10] a similar transformation did not occur on the ground. In short, the UGV era is still pending; their revolutionary potential remains latent.

There are good reasons to believe that UGVs will become a major – if not indispensable – feature of military land operations in the future. Broadly speaking, there is nothing inherently surprising in their military allure: they have the potential to save soldiers' lives – in essence, by putting metal before the flesh – and can make military missions more effective. Connected with this appeal, UGVs continue to produce high expectations about their near-term revolutionary impact on warfare. Yet it is not clear *a priori* whether these predictions will be realized.

In order to think more deeply about UGVs' likely revolutionary impact on future conflict, this article first provides a theoretical discussion about what makes any given technology revolutionary, in a military sense. Following on from this, it then details the developmental arc of UGVs against the backdrop of fluctuating military interest. The article then surveys current UGV programs, analyzing the technical and non-technical hurdles in the way of moving various types of UGVs from development into the field. It concludes by using the preceding analysis to evaluate the future impact of UGVs in ground combat.

As the question of UGVs' revolutionary potential is one part of a much larger debate about the role of robotics and autonomous systems and AI in future war, it is worth defining what I mean by a UGV before moving on to the analytical sections of this article.[11] A 'robotic' UGV is defined here as a mobile ground-based mechanical device that can sense and interact with its environment. This makes it distinct from an automaton that does the same thing every time once activated. Though varying in design, function, and means of navigation and control, most of the following components are typically common to most modern UGVs: *sensors* for the robot to perceive its surroundings, permitting controlled movement; a *platform* for locomotion, utility infrastructure and power to the system; usually an *affecter*, for manipulating the environment (this could be a weapon); a *control system*, the complexity depending on the level of autonomy[12]; the *human-machine interface*, which could be a joystick and a monitor control panel in the case of teleoperation, or something much more advanced; and a *communication link* for information exchange between humans and robots, and possibly between robots.[13]

The size and sophistication of UGVs, both armed and unarmed, vary considerably depending on their function, as does the level of autonomy. Typically, UGVs that are both lethal and autonomous are ascribed with the most potential to produce revolutionary change in military affairs. Before moving on to evaluate the future impact of UGVs as a whole, the following section provides a brief theoretical discussion about technology and military change.

Revolutionary military technology

The introduction of new military technology is especially important in theories concerning the sources of military revolutions, or major military innovations (MMI).[14] Indeed, technological change has accompanied many, but not all, the chief military innovations throughout history that have played a central role in determining the global balance of power.[15] Existing within a competitive environment, states seek to sustain or increase their relative power by improving upon their military capabilities.[16] A state's efficacy in harnessing technology is, *ceteris paribus*, highly connected with military prowess.[17] The possession of new technology is therefore intuitively attractive to military organizations charged with preparing for war and civilian leaders ultimately responsible for a state's survival.[18] It also stands to reason that the anticipated technological advances of adversaries induces fear.[19] 'Every country lives with the nightmare,' Henry Kissinger observed, 'that even if it puts forth its best efforts its survival may be jeopardized by a technological breakthrough on the part of its opponent.'[20] Unilateral possession of a potent new military technology – combined with its effective employment – can temporarily, yet decisively, upend a given balance of power.[21] States are attracted by the prospect of gaining advantages by generating new ways of producing military power. Although, as Michael Horowitz points out, 'inventing technologies or even being the first to use them does not guarantee advantages in international politics,'[22] in practice states always prefer to be the first to develop new technology.[23] The prospect of accruing military advantages through leveraging a new technology can lead to hype or technological exuberance. But how can we measure the potential of any given technology in its developmental infancy? Moreover, what constitutes *revolutionary* technology in military terms?

In his previous role as Secretary of the U.S. Army, Mark Esper opined, 'Whoever gets to robotics and AI first, it'll be a game changer on the battlefield.'[24] More particularly, UGVs have been touted as game-changing technology. This is particularly true of weaponized unmanned land systems rather than non-controversial, mundane uses of UGVs to enhance military effectiveness. But how much of this perceived revolutionary potential is hype rather than substance? Technological hype, along with synonym terms such as techno-optimism and technological exuberance,[25] denotes a condition whereby a technology generates excessive enthusiasm and high expectations.[26] International relations scholars and military historians have long observed, albeit episodically, that new technologies often trigger intense hype among civilian leaders, defense planners, and military officers.[27] Those enraptured by the emerging technology are pregnant with the belief that it will allow the user to either perform key established activities with increased efficiency or carry out entirely new activities altogether. However, it is not often clear which technologies will ultimately lead to military change, and, importantly, in what timeframe. Military thinkers have often peered into the future and got it wrong, misjudging the potential of technology or underestimating the time it can take for a technology to mature and emerge as something productive. Soon after the laser was first demonstrated by U.S. researchers in the early 1960s, it was considered by some as 'the biggest breakthrough in the weapons area since the atomic bomb.'[28] Yet more than half a century later, lasers are only just emerging as operationally useful weapons.[29]

Such technological hype may mask the many technical hurdles and adoption challenges that are often encountered in the process of maturing and fielding cutting-edge new weapons. As a consequence, disappointment with the technology can quickly follow when initial expectations are not met.[30] Forecasting the trajectory of new technology in its very early stages is especially difficult.[31] Cutting-edge technologies give rise to problems that are difficult to anticipate, understand, and to address.[32] In sum, making judgements about the future revolutionary impact of any given technology, and the pace at which is made productive, is intrinsically difficult. The remaining sections of this article use past and recent developments to better gauge how (and when) UGVs might substantially alter conflict on land.

The development arc of military UGVs

Military UGVs have a longer lineage than is often supposed. The Russian TT-26 *Teletank*, a small tracked device developed by the Red Army in the 1930s, and the *Goliath* fielded by Germany in WWII, could be loaded with explosives and controlled remotely from approximately a mile away.[33] Serving as single-use weapons, these tethered devices were more akin, however, to guided munitions than UGVs.[34] Ground robots – for military use or otherwise – really emerged in the second half of the twentieth century alongside major advances in computing.

With microprocessor technology flourishing in the 1970s, researchers in advanced nations envisioned robots replacing humans across a range of tasks in dangerous and hazardous environments, such as nuclear power plants[35] and in space.[36] Ground robots developed in this period were remotely controlled and equipped with sensors that fed back information to the operator via a console. Many key developments in robotics stemmed from U.S. defense research and development (R&D) funded projects in the 1980s, indicating robust military interest.[37] Prospective military uses focused on remote-controlled vehicles operating in chemical, biological, radiological, and nuclear (CBRN) environments or carrying out tasks that would otherwise expose vehicle occupants to excessive levels of danger, such as mine-clearance operations, the digging of tank firing pits, or rapid runway repair (RRR) under enemy fire. Other military applications that drew the attention of military planners, especially in the U.S., were for reconnaissance, surveillance, and target acquisition (RSTA)[38] missions and even launching weapons remotely.[39] Perhaps the most advanced American foray into the latter mission was the Tele-operated Mobile Anti-Armor Platform (TMAP) program, which envisaged remotely firing anti-tank missiles from a UGV.[40] This project, along with others, led to U.S. Congressional direction in December 1987 to prohibit the emplacement of weapons systems on robots.[41]

Despite dynamic technological developments beneficial to UGVs and robotics more generally in the 1990s – namely improvements in computer processing speed, reductions in the overall size and weight of electronics and sensors, the incorporation of semi-autonomous capabilities,[42] and widespread advances in communications – military interest waned at the close of the Cold War. It was not simply that the U.S. had lost a technological peer rival that had previously spurred competition in developing cutting-edge technologies; UGVs had not at this point established a track record of success. 'It isn't that the [robotic] systems weren't getting any better [in the 1990s],' Peter Singer argues, 'but that the interest, energy, and proven success stories necessary for them to take off just weren't there.'[43]

A more general optimism in robotics, however, continued apace in the new millennium. In 2002, then director of the MIT Artificial Intelligence Laboratory, Rodney Brooks, prophesized a robotics revolution was set 'to burst over us in the early part of the twenty-first century.'[44] Robotics enthusiasts forecast there would soon be a robot in every home and the family car would be self-driven.[45] Policymakers, especially in the U.S., regained interest in equipping militaries with uninhabited vehicles that were smaller (and cheaper), could provide greater endurance, and operate in more risky environments than traditional inhabited types. In 2001, the U.S. Congress issued a mandate that: 'It shall be a goal of the Armed Forces to achieve the fielding of unmanned, remotely controlled technology such that [...] by 2015, one-third of the operational ground combat vehicles are unmanned.'[46] UGVs subsequently were put front and center in the DoD's Future Combat Systems (FCS) project.[47] Initiated in 2003, and aimed to transform the army into smaller, lighter and more lethal units of manned and unmanned components,[48] the FCS program called for the extensive employment of UGVs, such as the MULE (Multifunction Logistics and Equipment Vehicle). Most significantly, the FCS program envisaged that, over the longer-term, the M1 Abrams main battle tank (MBT) and Bradley Fighting Vehicle could ultimately be replaced by remotely controlled robotic vehicles.[49]

The FCS project was overtaken by the events of 9/11 and the subsequent War on Terror. Counterinsurgency operations in Iraq and Afghanistan demanded the immediate fielding of small UGVs that many participating militaries did not possess. Small reconnaissance robots moving on treads or wheels and mounted with cameras and various sensors were rapidly procured off-the-shelf and integrated into U.S. and Coalition force structures. With IEDs the leading cause of casualties among U.S. forces in Iraq, small UGVs for EOD,[50] such as iRobot's *PackBot* (EOD version), which moved on tank tracks and had an extendable arm with high-powered zoom camera and claw-like gripper, became high demand items for the U.S. military.[51] Also, Coalition forces used variously named robots to 'search tunnels, caves, and buildings for enemy fighters and explosives.'[52] Early on, it seemed as though these counterinsurgency campaigns might also provide a testbed for lethal ground robots. In 2004 South Korea apparently sent two robot snipers to Erbil in northern Iraq. Later, the U.S. Army reportedly sent a few armed Talon SWORDS (Special Weapons Observation Reconnaissance Detection System) robots to Iraq.[53] Ultimately, lethal ground robots were not used.

Within a few years there was a menagerie of different robotic systems of all shapes and sizes deployed to Iraq and Afghanistan totaling in the thousands.[54] Ground robots deployed to Iraq and Afghanistan generated considerable excitement. Whilst these small, remotely controlled, battery-powered robots proved their worth for locating and examining objects, such as IEDs, the first decade of the new millennium did not usher in a new age of robotic warfare on the ground.[55] Work on a range of robotic prototype systems for other military tasks continued but often failed to pass muster during trails with potential end users.[56] The lack of development in larger unmanned ground platforms for use in more traditional conflict scenarios suffered as focus shifted to counterinsurgency operations after 2001. In the U.S. it was also a consequence of the DoD's decision, because of fiscal restraints (not least due to sequestration and the 2011 Budgetary Control Act), to cancel the flagship FCS program. With the return of great power competition, large UGVs – and especially armed varieties – are once again attracting considerable interest.

Great power competition and resurgent interest in UGVs

Resurgent enthusiasm in UGVs, and especially those for combat, has coincided with a return of great power competition. This is unsurprising. Logically, the allure of new military technologies heightens when strategic rivalry intensifies. Indeed, there is some historical evidence that large-scale geopolitical threats increase the urgency with which states respond to new military innovations, including the introduction of new technology.[57]

Unlike for much of the past half-century, the U.S. finds itself in a position of trying to play catchup when it comes to UGVs. Certainly it is fair to say that Washington can no longer assume it will always retain the technical edge in military robotics. Revisionist powers and even smaller states are seeking to capitalize on this technology.[58] The Russian Military Industrial Committee has purportedly approved a plan that would have 30 percent of Russian combat power consist of entirely robotic platforms by 2030.[59] This is aided by the fact that much of the cutting edge UGV technology in now developed by companies building systems for consumer markets, which have in turn flourished.[60]

With little UGV content harvestable from the 18 USD billion or so spent on the FCS program, the U.S. Army today lacks, outside of those for EOD operations, deployable unmanned ground capable of meeting the current or future operational requirements of a shifting international security environment in which the U.S. is striving to counter the emerging military capabilities of near-peer competitors such as Russia and China.[61] This is despite year-on-year increases in past three budgets. Paul Scharre argues: 'what may look on the outside like an unmitigated rush toward robotic weapons is, in actuality, a much more muddled picture inside the pentagon.'[62] DoD funding on UGVs has been consistently below five percent of total unmanned systems spending.

The U.S. Army has tried to shape the American ground robotics industry, launching competitions to define its future unmanned ground systems fleet.[63] It continues to trial systems, such as the squad mission support system (SMSS), an unmanned vehicle primarily designed for logistics support for light and early-entry forces.[64] Furthermore, there are enduring plans to reduce brigade combat team size by a quarter (from 4,000 to 3,000 troops) by replacing traditional supply convoys with trains of semi-autonomous robot vehicles.[65] Yet serious questions about the U.S. commitment to UGVs continue and have grown louder with the activities of America's near-peer competitors in this area. Reflecting this, *The U.S. Army Robotics and Autonomous Systems Strategy* published in 2017 sets out as a mid-term priority (2021–2030) the need to have full automated convoy operations and unmanned combat vehicles improving maneuver.[66]

Russia's foray into armed UGVs has been a major catalyst for widespread renewed interest in unmanned ground systems. Weaponized systems such as the Uran-9, armed with a 7.62-mm machine gun, a 30-mm automatic cannon and four anti-tank missiles and designed to provide combat units with remote surveillance and fire support, caused major alarm in many capitals when news emerged that they were deployed and perhaps tested in 'combat conditions' in Syria 2015.[67] More recently, the manufacturer of the *Soratnik* automated fighting vehicle claims to have tested the system in conditions maximally close to a real combat environment.[68] Russia has unveiled other combat UGVs, like the *Vikhr* (Whirlwind) which is based on its BMP-3 infantry fighting vehicle.[69] Able to accommodate a variety of weapons, the *Vikhr* is said to be remotely controlled up to a distance of 10 kilometers.

Whilst Russia may be testing a variety of unmanned systems for combat, developing test beds and putting them through robust trials, 'Russian military manufacturers and contractors are,' Samuel Bennet, an expert on Russia's UGVs, reminds us, 'not that different from their American counterparts in sometimes talking up the capabilities of their creations, seeking to create the demand for their newest achievement before there is proof that such technology can stand up to harsh battlefield conditions.'[70] Furthermore, boastful comments by Russian officials, such as those by the Russian Defense Minister Sergei Shoigu in 2018 stating that the serial production of ground combat robots for the military 'may start already this year [2018],'[71] do not mean fleets of armed UGVs are imminent. Tellingly, the Russian military have not received these systems into their own service yet, and the technology is some way away from export readiness. Nonetheless, Russia's attempt to trial weaponized UGVs with automated features has amplified attention.

As for China, according to the U.S. Defense Science Board, 'every major manufacturer for the Chinese military has a research center devoted to unmanned systems.'[72] China North Industries Corporation (Norinco), the primary supplier of land combat platforms and weapons to the People's Liberation Army (PLA), has signaled its intent to increase its presence in the unmanned ground vehicle (UGV) market. It recently unveiled four new prototype unmanned combat and logistics UGVs as well as updates to two of its existing vehicles.[73] Some analysts believe that it is working on even more cutting-edge unmanned systems.[74]

Beyond the great powers, a growing number of nations are embarking on UGVs that can be weaponized. The British Army recently tested the *Titan*

Sentry and Strike, a UGV developed by QinetiQ and Estonia's Milrem Robotics. It can be used for carrying large payloads, as a medevac or fitted with a machine gun for attack roles.[75] Germany's Rheinmetall has developed the *Mission Master*, a multimission Unmanned Ground Vehicle (UGV) that can perform various mission profiles, from mule to force protection and surveillance, using a variety of modular payloads.[76] Even countries not considered technology leaders are producing combat varieties of UGVs. The *Nazir* entered service with the Iranian military in 2016, and is reportedly capable of mounting a variety of missiles or machine guns.[77] Even Shia militias in Iraq have gotten in on the game, fielding an armed ground robot in 2015.[78]

Robotics technology is increasingly cheap and mass producible – it does not require a massive industrial complex, not only to build but to operate. Many of the component items if not whole systems can now be purchased commercially. UGVs are arguably reaching a stage of maturity whereby they will make a significant debut in ground operations. At the very least, it seems that various governments are banking on such systems soon playing a major if not revolutionary role. Despite renewed hype about large UGVs, are there reasons for a more cautious assessment about their impending revolutionary impact?

Impediments to near-term adoption

When thinking about the potential barriers to an 'unmanned' future, policymakers and academics have identified several organizational impediments.[79] Military organizations typically face many adoption challenges related to adapting doctrine, tactics, and infrastructure support, in fielding new weapons.[80] Before UGVs can be used, users must develop workable operational concepts.[81] Organizational cultural reasons may make some militaries reluctant to switch to unmanned vehicles.[82] Even the conversion of supply trucks to unmanned vehicles was not enacted by the U.S. Army despite the availability of the kits and the lives that might have been saved.

For many, the technological case for military consumers to adopt UGVs is compelling. Their efficacy lies in the potential for reducing manpower requirements as the robot assumes human tasks or functions. This has the potential for less warfighters being sent into combat, as robots take up positions and conduct tactical tasks. Early manpower assessments regarding UGV capabilities came to the conclusion that one individual could be trained to operate multiple tele-operated systems. This assessment, however, was later proven inaccurate.[83] Regardless of a new weapon's technological merits, Christopher Coker observes that: 'We make the mistake of centering our histories of technology on *innovation* rather than *use*.'[84] This bias occurs because all too often the conventional technological life cycle has a producer-orientated approach. Within the scholarship on unmanned technologies there is increasingly interest in the importance of experiential theories of technological development and end-user adoption.[85] As Eliot A. Cohen notes, technology alone 'does not yield transformational change, particularly in the military world'; indeed a 'lag between the appearance of useful technologies and their mature application in organizations and concepts of operations is not surprising. What is surprising is ... the relative paucity of attention to the human software, as it were, as opposed to the material.'[86] Recognizing this as an impediment, the U.S. military has fielded testbed vehicles to allow users to develop and refine UGV operational concepts.[87] As UGVs are increasingly socialized among military constituents, organizational barriers will lower. Moreover, militaries typically emulate the successes of others. Once one military has successfully debuted UGVs operations, others will attempt to copy this if they can.

Some argue that ethical and legal considerations will halt the introduction of more autonomous robots. UGVs for military applications such as logistics and surveillance are unlikely to raise appreciable objections. But whether autonomous and weaponized UGVs should be permitted to take life is another matter. Battlefield pressures could drive militaries to build autonomous weapons that take the humans out of the loop all the same.[88] Normative resistance will likely crumble if rivals are perceived to have gained advantages from the technology. At any rate, although political constraints in some jurisdictions have acted as a brake on the prospective use of ground variants of lethal autonomous weapon systems (LAWS), this fails to account for why so few uncontroversial vehicular military tasks are being performed by habited vehicles.

As for technical hurdles, it is not certain that the technology is developing in some key areas as rapidly as is the hype about UGVs. As one technical report notes: 'Developing autonomous robotic ground systems that can successfully navigate tactically cross-country is a significant challenge that will need to be overcome before these systems can be employed effectively on the battlefield.'[89] Navigation in cluttered, rocky, or obstacle-dense urban environments is very difficult.[90] With tele-operated UGVs, all too often the operator misjudges the proximity of obstacles due to the lack of depth perception on the OCU monitor. This has not been so much of a problem for UGVs used for countering IEDs, which require slow, deliberate movement.[91] Without high levels of mobility, UGVs cannot participate in high tempo, fluid operations. Indicative of this is the fact that the most prevalent operational use of armed UGVs is for sentry duties, such as the Russian *Taifun-M*, which features laser targeting and a cannon and is reputedly being used as a remote sentry to guard Russian missile sites.[92] In another example, Israel has a developed a sentry UGV to patrol parts of its border. The site security (sentry) application matches the strengths and avoids the weaknesses of UGVs. The operating environment is known in advance, is under friendly control, and can to some degree be tailored to support robot operations.

To highlight some of the enduring technical challenges, it is worth focusing on the most prominent recent case of a combat UGV tested in an operational theatre – that of the Russian Uran-9 in Syria.[93] Post-operational reporting by Russian researchers noted that the tank's controllers unexpectedly lost control of the Uran-9 on 17 occasions for short periods and for over an hour on two further occasions.[94] Also, the system was unable to operate far from its controllers,[95] especially in built-up areas, and encountered difficulties firing its 30 mm gun, including an inability to fire on the move, wiping out its very purpose as a 'combat' UGV.[96] The electro-optic stations allowed for reconnaissance and identification of potential targets at a range of no more than two kilometers. Moreover, it did not allow for adequate detection of adversary's optical and targeting devices and created multiple interferences in the test range's ground and airspace.

But many of the technological limitations reported about the Uran-9 are also found across the global development of unmanned systems. Communication is essential in the case of military robots, where both accuracy and secrecy of information exchange are crucial.[97] Yet it is still proving difficult to build untethered systems with solid communications.[98] What's more, non-line of sight or beyond line of sight (BLOS) operation is limited to a UGV's ability to send and receive communications.[99] Radio frequency and electro-magnetic interference are prevalent on the ground and cause signal degradation during UGV tele-operation.[100] To date, UGVs have operated in a relatively permissive electromagnetic environment (such as for U.S. forces in Iraq and Afghanistan). The electromagnetic spectrum will be highly contested in the future, creating major communication vulnerabilities between UGVs that are not fully autonomous and their controllers.

The final acceptance of UGV applications will also depend on their reliability. As Jennifer Carlson and Robin Murphy predict: 'A fragile robot in constant need of maintenance and repair is likely to be left behind to make room for more reliable equipment.'[101] Despite their overall success, small tele-operated devices in Iraq and Afghanistan in EOD and reconnaissance roles, encountered a reliability problem in the field, apart from lab conditions.[102] Common mechanical failures ranged from broken shear pins to thrown tracks. Dirt simply entering the moving devices was often enough to cause wear and eventually critical faults. As for sensors, bumpy terrain, sudden changes in lighting, and rainy weather caused problems for the on-board cameras, making it difficult to remotely control the robot. A further persistent common failure with many UGVs was with their control systems, leading to sluggishness in movement, a complete loss of steering, or the robot becoming entirely unresponsive.[103] Dropped signals in wireless controlled robots were the most frequently encountered problems in communication.[104] These reliability issues have not been ironed out.

One way to overcome many of these limitations is to build autonomous UGVs that can self-regulate and operate without the need for a human controller.[105] For one thing, providing the system with AI to execute tasks on its own ensures limited or no loss of the UGV from communications interference. Greater autonomy would give the UGV the ability to gain information about the environment and to work for extended durations without human intervention. To act as independent fighters they will need to move autonomously. That means they must have sophisticated vision systems that not only see what is around them but also *interpret* what they see.[106] This is essential for uncontrolled environments where navigation around obstacles, even at low speeds, is difficult.[107]

None of the factors discussed above will likely halt the eventual mass introduction of all kinds of UGVs, including weaponized systems with increasing autonomy. But many of these abovementioned impediments are all too often missing in discussions about the future unmanned battlefield. Past experience tells us that technologies advance in fits and starts – not a steady upward curve.

Conclusion: an impending UGV revolution in ground operations?

Whilst enthusiasm about a technology's ability to transform established ways of doing things in the short-term is often be misplaced, it is not necessarily ill-founded when viewed over the long-term. Francis Collins, the American physician-geneticist who led the Human Genome Project, called the overestimation of the short-term impact of a truly transformational discovery the 'First Law of Technology.'[108] Likewise, Futurist Paul Saffo has termed the overestimation of a new technology's short-term impact and failure to anticipate longer-term implications after initial inflated expectations are dashed as 'macro-myopia.'[109]

With those technologies that eventually mature their attraction increasingly stems from their performance attributes rather than simply their promise.[110] And as the technology becomes valued by end-users, adoption accelerates. Nonetheless, it is misguided, as Christopher Coker notes, 'to date advancement and progress from the moment a technology appears or is first applied, and to downplay the long and winding road of adoption, imitation, diffusion, improvement, recycling and hybridization.'[111] There are early signs that UGVs have been valued when they have proved that they can perform tasks more effectively than using traditional equipment or have been shown to have reduced danger to soldiers' lives. Iraq – and to a lesser extent Afghanistan – changed end-user perception about ground robots. Here robots offered a potential solution to a problem that has bedeviled counter-insurgents – public's lack of tolerance for spilt blood in military campaigns where the homeland is not attacked.

In this way, UGVs have to date been a sustaining technological innovation, that is, they have improved performance along established trajectories familiar to traditional customers.[112] It is not yet clear whether UGVs will emerge as a disruptive technological innovation, producing new performance metrics altogether.[113] The extent to which UGVs will revolutionize military affairs will depend much on the operational concepts militaries will develop for their employment. Early ideas in Russian thinking have armed UGVs storming stationary and well-defended targets (effectively giving such combat UGVs a kamikaze role) followed up by more conventional forces. Others foresee the future battlefield essentially depopulated by humans and instead monopolized by LAWS.

Military robot developments are progressing on many fronts and are underpinned by a diversity of technologies such as artificial intelligence and advanced computation, sensors and vision systems, new materials, guidance and location techniques, power sources and propulsion technologies. As machine intelligence advances, states will field increasingly autonomous robots capable of carrying out more complex tasks. Even so, for the next 10–15 years, it is unlikely that UGVs will be able to effectively take part in complex, fast-paced tasks, especially combat. There remain significant challenges even for the most robust of ground-based platforms. Overcoming such challenges will be expensive in terms of cost and development time.

Notes

1. Scharre, *Army of None*; Fielding, "Robotics in Future Land Warfare"; and Sloan, "Robotics at War."
2. Congressional Research Service, *U.S. Ground Forces Robotics*.
3. See in particular Horowitz, Kreps, and Fuhrmann, "Separating Fact from Fiction."
4. The academic community usually refers to UGVs (especially UGVs possessing significant autonomous capabilities) simply as mobile robots. This article uses both terms interchangeably. In this study, UGVs do not include bipedal humanoid types of robots or powered exoskeletons.
5. Shaker and Wise, *War without Men*.
6. Bode and Huelss, "Autonomous Weapons Systems."
7. Much recent commentary about UGVs focuses on the future role of ground variants of lethal autonomous weapon systems (LAWS), especially in matters related to the legality and ethics of their use. Refer to Klare, "Autonomous Weapons Systems and the Laws of War."
8. Cited in: https://vpk-news.ru/sites/default/files/pdf/VPK_08_476.pdf. For a translation look to Coalson, "Top Russian General Lays Bare."
9. Krieg and Rickli, *Surrogate Warfare*; and Rossiter, "Participation in Warfare." On the opposing side, insurgents have also looked to drones to increase the distance between attackers and attacked. See Rossiter, "Drone Usage by Militant Groups."
10. Hall, "Drones: Public Interest, Public Choice."
11. Rossiter, "Robotics, Autonomous Systems, and Warfare."

12. For an accessible primer on this topic, see Murphy, *Introduction to AI Robotics*. For teleoperated UGVs, externally situated humans transmit navigational guidance to the vehicle. Autonomous vehicles determine their own course using onboard sensors and processing capabilities. The term supervisory control is often given to control schemes that combine inputs from external human operators and onboard sensors to determine the navigation.
13. Underpinning all of this is system integration, which provides synergy within a robotic system.
14. Gray, *Strategy for Chaos*; Boot, *War Made New*; Krepinevich, "Cavalry to Computer"; Van Creveld, *The Transformation of War*; and Mahnken, *Technology and the American Way of War*.
15. RMAs or MMIs can also of course be driven by major organizational changes to militaries. See Horowitz, *The Diffusion of Military Power*, 22.
16. Fastabend and Simpson, "Adapt or Die."
17. McNeill, *The Pursuit of Power*. Whilst other analysts consider the effectiveness of force employment and relative skill levels of military personnel important determiners of military potency, few would disagree that technological superiority is typically an advantage. On the importance of force employment and skill levels in producing military effectiveness, see in particular Biddle, *Military Power*.
18. An imprecise concept, technology can refer to products, to technology and skills or the activities in developing technologies. This analysis employs the more limited sense of the term of technology as a product, i.e. as a weapon, in a military context. Although the phrase 'military technology' is used throughout, the study recognizes that the origins may be either civilian or defense. Smit, "Science, Technology and the Military."
19. For the effects of the Sputnik launch on U.S. policy, refer to Peoples, "Sputnik and 'Skill Thinking' Revisited."
20. Kissinger, "Arms Control, Inspection and Surprise Attack."
21. A number of studies have found that sudden shifts in the military balance between dyads – especially those with a history of animosity – raises the likelihood for war. For synoptic coverage of work on this theory, refer to Geller and Singer, *Nations at War*, 147. See also Gilpin, *War and Change in World Politics*, 59–62; and Powell, *In the Shadow of Power*.
22. Horowitz, *The Diffusion of Military Power*, 2. Indeed, there are some historic examples of states learning from and improving upon the efforts of first mover that produce better long term results. Japan and the U.S. accrued greater benefits from carrier warfare even though Britain was the first mover. See Hone, Friedman, and Mandeles, *American and British Aircraft Carrier Development*.
23. Mearsheimer, *The Tragedy of Great Power Politics*, 232.
24. Freedberg Jr., "Evolutionary Revolution."
25. This term was coined by York in *The Advisers*, ix.
26. Van Lente, Spitters, and Peine, "Comparing Technological Hype Cycles."
27. Mahnken, *Technology and the American Way of War*; Fino, "Breaking the Trance"; Howard, "How much can Technology Change Warfare"; Freedman, *The Future of War*; and Rossiter, "High-energy Laser Weapons."
28. Letter from Major General A. Schomburg to Lieutenant General J. H. Hinrichs, 16 January 1962, history office, US Army Missile Command, Redstone Arsenal, Huntsville, AL, quoted in Seidel, "From Glow to Flow," 114.
29. See Rossiter, "High-energy Laser Weapons."

30. This at least seems to be the case with the ABL project. The program was cancelled in 2012 after encountering insurmountable technical obstacles and rising costs.
31. Dedehayir and Steinert, "The Hype Cycle Model," 28. On this point see again Lente, *Promising Technology*; and Guice, "Designing the Future."
32. Von Hippel, "'Sticky Information' and the Locus of Problem Solving."
33. Two battalions of the Teletanks were deployed in the Winter War against Finland in 1940.
34. Michel and Gettinger, "Out of the Shadows."
35. Tsitsimpelis Taylor, Lennox, Joyce, "A Review of Ground-based Robotic Systems."
36. In 1976, robots were put onboard the Viking 1 and 2 space probes.
37. Gage, "A Brief History of Unmanned Ground Vehicle (UGV)."
38. A U.S. Marine Corps' UGV RSTA program in the early 1980 s called the Ground Surveillance Robot (GSR), for example, successfully tested, using a fully actuated 7-ton M-114 armored personnel carrier as the testbed host vehicle, the ability to follow a lead vehicle autonomously. See: Harmon, "The Ground Surveillance Robot (GSR)."
39. The U.S. Army and the U.S. Marine Corps each pursued a number of battlefield (RSTA and weapons-launching) UGV developments in the 1980s.
40. Prototype systems were even procured in 1987/1988 from Grumman and Martin Marietta. Both systems were joystick-controlled via fiber optic link, the operator navigating via the returned TV image. The Grumman system was a hybrid diesel-electric drive with its four wheels in an articulated diamond pattern, while the Martin vehicle was a diesel-powered hydrostatic four-wheel drive with skid steering.
41. Consequentially, the TMAP was retargeted to the RSTA mission. It was renamed the Teleoperated Mobile All-Purpose Platform. Refer to Young, "Military Explores Robotic Technology."
42. Arguably, progress in mobile ground robots is inextricably linked to AI. Because much of modern UGV research has been related to mobility, developments in artificial intelligence (AI) and automation was behind much of the rapid capability gains. The development of autonomous mobile robots with nontrivial navigational capabilities began as an interesting application domain for Artificial Intelligence researchers in the late 1960s. See DARPA, *Strategic Computing Program Second Annual Report*.
43. Singer, *Wired for War*, 54.
44. Brooks, *Flesh and Machines*, 6 and 10–11.
45. On technological exuberance and robotics at the turn of the century, see Singer, *Wired for War*, 8.
46. National Defense Authorization Act for Fiscal Year 2001, US Public Law 106–398, Section 220, 106th US Congress, 2nd session, 2000, accessed at: https://www.govinfo.gov/content/pkg/PLAW-106publ398/pdf/PLAW-106publ398.pdf.
47. Scharre, *Army of None*, 61.
48. See the US Congressional Budget Office report: The Army's Future Combat Systems Program and Alternatives (August 2006), https://www.cbo.gov/sites/default/files/cbofiles/ftpdocs/74xx/doc7461/08-02-army.pdf.
49. The Armed Robotic Vehicle, a five-ton mini-tank that could be equipped with missiles or a.30 mm chain gun; and the Soldier Unmanned Ground Vehicle,

a 30-pound, man-portable scout equipped with weapons and sensors. Refer to Fish, "UGVs in Future Combat Systems."

50. Scharre, *Army of None*, 14. Britain was perhaps the first to field a modern military UGV: the *Morfax Wheelbarrow*, a remote controlled Explosive Ordinance Disposal (EOD) unit deployed to Northern Ireland in the early 1970s.
51. As did the *Talon*, a two-foot-six-inch robot which looks like a miniature tank made by Foster-Miller. By 2008 there were close to 2,000 *Talons* in the field and the firm had won a $400 million contract to supply another 2,000. Singer, *Wired for War*, 23 and 26–27.
52. Boot, 'The Paradox of Military Technology,' 24.
53. These can be equipped with M240 or M249 machine guns, Barrett.50 caliber rifles, 40 mm grenade launchers, or antitank rocket launchers, SWORDS were controlled remotely by a soldier using a video screen and joystick. Refer to M. E. Purdy, Presentation slide "Ground robotics technology," Joint Ground Robotics Enterprise, Department of Defense, June 2007.
54. Morris, "Military Projects 4,000 Robots," 4; and Bachman, "The U.S. Army is Turning to Robot Soldiers."
55. Sharkey, "Cassandra or False Prophet of Doom."
56. A good example is *BigDog*, a dynamically stable quadruped military robot created in 2005 by Boston Dynamics with Foster-Miller that received considerable publicity but was ultimately disappointment. Hern, "US Marines Reject BigDog."
57. Posen, *The Sources of Military Doctrine*: 239–40.
58. Congressional Research Service, *U.S. Ground Forces Robotics and Autonomous Systems (RAS)*, 5.
59. Mike Ryan, *Human-Machine Teaming*, 13.
60. Global spending on robotics and UAVs reached USD116 billion last year, with a compound annual growth rate (CAGR) of up to 20 percent, pushing spending to a predicted USD210 billion by 2022. As a result, there is a widening capability gap between commercial and military systems.
61. Summary of the *2018 National Defense Strategy of the United States of America: Sharpening the American Military's Competitive Edge*, 7. Accessed at: https://dod.defense.gov/Portals/1/Documents/pubs/2018-National-Defense-Strategy-Summary.pdf.
62. Scharre, *Army of None*, 61.
63. Judson, "Army Poised to Transform Ground Robotics Industry."
64. Refer to https://www.army-technology.com/projects/squad-mission-support-system-smss/.
65. Atherton, "Robots May Replace One-Fourth"; and Farmer, "US Army considers replacing thousands."
66. See the U.S. Army Training and Doctrine Command, *The U.S. Army Robotic and Autonomous Systems Strategy*, March 2017, 5. Accessed at http://www.arcic.army.mil/App_Documents/RAS_Strategy.pdf.
67. According to statements published in Russia's state-owned RIA Novosti, the Russian defense ministry announced that, as well as the Uran-6 designed for mine clearance, the Uran-9 multifunctional reconnaissance and fire support system was deployed in Syria, though this has not been verified. See Atherton, "Russia Confirms its Armed Robot Tank was in Syria." There is no hard evidence that the Russians have tested unmanned ground vehicles in actual combat other than relatively basic explosive ordnance disposal/mine-clearing robots. A video of the robot can be viewed at: www.youtube.com/watch?vKOZbXwjwoOg.

68. See: "Russia tests robotic strike vehicle in conditions close to real combat," *Tass News Agency*, 19 January 2018, https://tass.com/defense/985821.
69. Novichkov, "New Russian Combat UGV Breaks Cover," 30.
70. Bendett, "Russian Ground Battlefield Robots."
71. Refer to "Combat robots for Russian troops to go into serial production this year – defense minister," *Tass News Agency*, 15 March 2018, https://tass.com/defense/994310.
72. See http://www.acq.osd.mil/dsb/reports/AutonomyReport.pdf.
73. Wong, "China's Norinco rolls out new combat."
74. Dominguez, "China Seeking to Surpass U.S.," 6.
75. Husseini, "In Pictures: The UK MoD's Future Robot Army."
76. See: https://www.rheinmetalldefence.com/en/rheinmetall_defence/systems_and_products/unbemannte_fahrzeuge/index.php.
77. A larger version has recently been unveiled by Iran. See Atherton, "Did Iran just show off a new ground robot?"
78. Scharre, *Army of None*, 102.
79. See for example, Horowitz, *Diffusion of Military Power*.
80. Gilli, and Gilli, "The Diffusion of Drone Warfare?"
81. Ryan, *Human-Machine Teaming*, 38.
82. One impediment for the use of Unmanned Combat Aerial Vehicles (UCAVs) is the resistance of pilots whose professional standing is linked to piloted planes.
83. The minimum ratio for operator-UGV control is 1:1 as seen in many of the RSTA-specific UGVs used in Iraq and Afghanistan. For larger, more complex systems that incorporate complicated EO/IR suites or lethal payloads, the ratio jumps to 2:1 and 3:1. Evans III and Jentsch, "The Future of HRI," 435.
84. Coker, *Future War*, 17. Original italics.
85. Gilli and Gilli, "The Diffusion of Drone Warfare?"; Fuhrmann and Horowitz, "Droning On"; Horowitz, Kreps, and Fuhrmann, "Separating Fact from Fiction"; and Grissom, "The Future of Military Innovation Studies."
86. Cohen, "Change and Transformation in Military Affairs," 399. For similar points, see Farrell, Osinga, and Russell, (eds.), *Military Adaptation in Afghanistan*, 10; Murray, *Military Adaptation in War*, 8–13 and 15.
87. The idea has been to put prototype hardware into the hands of prospective users to conduct operational appraisals while using the systems, creating a sense of ownership among the user community, as well as provide constructive feedback to the developers.
88. Scharre, *Army of None*, 8.
89. Congressional Research Service, *U.S. Ground Forces Robotics and Autonomous Systems (RAS)*.
90. This is partly due to the fact that traditional tracked and wheeled robots must reorient to perform some maneuvers, such as lateral displacement. See Iagnemma, Shimoda, and Shiller, "Near-optimal Navigation."
91. Elliot Redden, "Robotic Control Systems for Dismounted Soldier," 346.
92. It is fitted with a mast-mounted radar and infrared (IR) and optical sensors, and some sources report that it can locate enemy vehicles at a range of 6 km and troops at 3 km. A video clip showing the unit is available at: www.youtube.com/watch?vuFC4yMMtHG0.
93. Much of the following information is derived from Andrei P. Anisimov, Senior Research Officer at the 3rd Central Research Institute of the Ministry of Defense. His reported on the Uran-9's critical combat deficiencies during the 10th All-Russian

Scientific Conference entitled "Actual Problems of Defense and Security," held in April 2018.

94. See a translation of key parts of the presentation delivered by a senior Russian Ministry of Defense scientific researcher at a 2018 security conference at the Naval Academy in St Petersburg at: https://defence-blog.com/army/combat-tests-syria-brought-light-deficiencies-russian-unmanned-mini-tank.html.
95. Instead of its intended range of several kilometers, the Uran-9 could only be operated at distance of 300–500 meters among low-rise buildings, wiping out up to nine-tenths of its total operational range.
96. Brown, "Russia's Uran-9 Robot Tank"; and Mizokami, "Russia's Tank Drone."
97. One of the reasons UAS technologies have been more aggressive in their development than the UGV is the fact that UAS communication's architecture, including command and control nodes, pre-existed and the maneuver space aircraft use contains less obstructions to the command and control links from Ground Control Station to the UAS.
98. Minkov and Oron-Gilad, "Remotely Operated Vehicles (ROVs)," 211.
99. Zhang et al., "Unmanned Ground Vehicle Navigation."
100. Lee, "Future Unmanned System Design," 177.
101. Carlson and Murphy, "How UGVs Physically Fail," 424.
102. Reliability studies demonstrated that the Mean Time between Failure (MTBF) for the types of UGVs deployed to Iraq was far too low. Ibid.; and Smuda et al, "Deploying of Omni-directional Inspection System."
103. Tiron, "Ground Robotics Experience Bumpy Ride."
104. Nguyen-Huu et al., "Reliability and Failure in Unmanned"; Kenyon, "U.S. Robots Surge onto the Battlefield."
105. This is not new. AI has always been at the center of mobile ground robot research. See Klafter, "Mobile Robots, Research and Development."
106. Lamb, "Battle of the Bot."
107. At present, humans remain better drivers on snow-covered roads, driving into the sun, and driving in rain or dust storms than robots.
108. Collins, "A Genome Story."
109. See transcript of PBS interview with Paul Saffo at: https://www.pbs.org/wgbh/pages/frontline/cyberspace/saffo.html.
110. Dosi, "Technological Paradigms and Technological Trajectories"; and Jun, "A Comparative Study of Hype Cycles, 1414.
111. Coker, *Future War*, 17.
112. Dombrowski and Gholz, "Identifying Disruptive Innovation," 104.
113. Christensen, *The Innovator's Dilemma*, xv.

Disclosure statement

No potential conflict of interest was reported by the authors.

ORCID

Ash Rossiter http://orcid.org/0000-0001-5427-2229

Bibliography

Atherton, Kelsey D. 2019. "Did Iran Just Show off a New Ground Robot?" *C4ISRNET*, February 1. https://www.c4isrnet.com/unmanned/2019/02/01/did-iran-just-show-off-a-new-ground-robot/.

Atherton, Kesley D. 2014. "Robots May Replace One-Fourth of U.S. Combat Soldiers by 2030, Says General." *Popular Science*, January 22. https://www.popsci.com/article/technology/robots-may-replace-one-fourth-us-combat-soldiers-2030-says-general/.

Atherton, Kesley D. 2018. "Russia Confirms Its Armed Robot Tank Was in Syria." *Defense News*, May 7. https://www.defensenews.com/unmanned/2018/05/07/russia-confirms-its-armed-robot-tank-was-in-syria/.

Bachman, Justin. 2018. "The U.S. Army Is Turning to Robot Soldiers." *Bloomberg*, May 18. https://www.bloomberg.com/news/articles/2018-05-18/the-u-s-army-is-turning-to-robot-soldiers.

Bendett, Samuel. 2018. "Russian Ground Battlefield Robots: A Candid Evaluation and Ways Forward." *Blog on Mad Scientist Laboratory*, June 25. https://madsciblog.tradoc.army.mil/63-russian-ground-battlefield-robots-a-candid-evaluation-and-ways-forward/.

Biddle, Stephen. *Military Power: Explaining Victory and Defeat in Modern Battle*. Princeton, NJ: Princeton University Press, 2004.

Bode, Ingvild, and Hendrik Huelss. "Autonomous Weapons Systems and Changing Norms in International Relations." *Review of International Studies* 44, no. 3 (2018): 393–413. doi:10.1017/S0260210517000614.

Boot, Max. Fall 2006. "The Paradox of Military Technology." *The New Atlantis*, 13–31.

Boot, Max. *War Made New: Technology, Warfare, and the Course of History: 1500 to Today*. New York: Gotham Books, 2006.

Brooks, Rodney A. *Flesh and Machines: How Robots Will Change Us*. New York: Pantheon Books, 2002.

Brown, Daniel. "Russia's Uran-9 Robot Tank Reportedly Performed Terribly in Syria." *Business Insider* 9, no. July (2018): 9. https://www.businessinsider.com/russias-uran-9-robot-tank-performed-horribly-in-syria-2018-7.

Carlson, Jennifer, and R.R. Murphy. "How UGVs Physically Fail in the Field." *IEEE Transactions on Robotics* 21, no. 3 June (2005): 423–437. doi:10.1109/TRO.2004.838027.

Christensen, Clayton M. *The Innovator's Dilemma: When Technologies Cause Great Firms to Fail*. Boston: Harvard Business School Press, 1997.

Coalson, Robert. 2014. "Top Russian General Lays Bare Putin's Plan for Ukraine." *Huffington Post*, September 2, https://www.huffpost.com/entry/valery-gerasimov-putin-ukraine_b_5748480.

Cohen, E A. "Change and Transformation in Military Affairs." *Journal of Strategic Studies* 27, no. 3 (2004): 395–407. doi:10.1080/1362369042000283958.

Coker, Christopher. *Future War*. Cambridge, UK: Polity Press, 2015.

Collins, Francis. 2010. "A Genome Story: 10th Anniversary Commentary by Francis Collins." *Blog in Scientific American*, June 25. https://blogs.scientificamerican.com/guest-blog/a-genome-story-10th-anniversary-commentary-by-francis-collins/.

Congressional Research Service. 2018. *U.S. Ground Forces Robotics and Autonomous Systems (RAS) and Artificial Intelligence (AI): Considerations for Congress*. Updated November 20 https://crsreports.congress.gov/product/pdf/R/R45392.

Dedehayir, Ozgur and Martin Steinert. "The Hype Cycle Model: A Review and Future Directions." *Technological Forecasting and Social Change* 108 (2016): 28–41.

Defense Advanced Research Projects Agency (DARPA). *Strategic Computing Program Second Annual Report*. Arlington, VA (1986).

Dombrowski, Peter, and Eugene Gholz. "Identifying Disruptive Innovation: Innovation Theory and the Defense Industry." *Innovations: Technology, Governance, Globalization* 4, no. 2 (2009): 101–117. doi:10.1162/itgg.2009.4.2.101.

Dominguez, Gabriel. 2018. "China Seeking to Surpass U.S. In Military Technologies, Says Canadian Report." *Jane's Defence Weekly*, June 6.

Dosi, G. "Technological Paradigms and Technological Trajectories." *Research Policy* 11, no. 3 (1982): 147–162. doi:10.1016/0048-7333(82)90016-6.

Evans, William A., III, and Florian G. Jentsch. "The Future of HRI: Alternate Research Trajectories and Their Influence on the Future of Unmanned Systems." In *HumanRobotic Interactions in Future Military Operations*, edited by Michael Barnes and Florian Jentsch, 435–442. Surrey, UK: Ashgate Publishing Limited, 2010.

Farmer, Ben. 2014. "US Army Considers Replacing Thousands of Troops with Robots." *Daily Telegraph*, January 21. https://www.telegraph.co.uk/news/uknews/defence/10587820/US-Army-considers-replacing-thousands-of-troops-with-robots.html.

Farrell, Theo, Frans Osinga, and James A. Russell, eds. *Military Adaptation in Afghanistan*. Stanford, CA: Stanford University Press, 2013.

Fastabend, David A., and Robert H. Simpson. "Adapt or Die." *Army* (February 2004).

Fielding, Marcus. "Robotics in Future Land Warfare." *Australian Army Journal* 3, no. 2 (2006): 99–108.

Fino, Steven A. "Breaking the Trance: The Perils of Technological Exuberance in the U.S. Air Force Entering Vietnam." *Journal of Military History* 77 (2013): 625–655.

Fish, Scott. "UGVs in Future Combat Systems." In *Unmanned Ground Vehicle Technology VI, Proceedings of SPIE*, edited by Grant R. Gerhart, Chuck M. Shoemaker, and Douglas W. Gage. International Society for Optics and Photonics 5422, 2004.

Freedberg, Sydney J., Jr. 2018. "Evolutionary Revolution: Esper's Army Modernization Philosophy." *BreakingDefense.com*. May 16. https://breakingdefense.com/2018/05/evolutionary-revolution-sec-espers-army-modernization-philosophy/.

Freedman, Lawrence. *The Future of War: A History*. London: Penguin Books, 2018.

Fuhrmann, Matthew, and M C. Horowitz. "Droning On: Explaining the Proliferation of Unmanned Aerial Vehicles." *International Organization* 71, no. 2 (2017): 397–418. doi:10.1017/S0020818317000121.

Gage, Douglas W. 1995. "A Brief History of Unmanned Ground Vehicle (UGV) Development Efforts." [Special Issue on Unmanned Ground Vehicles] *Unmanned Systems Magazine*.

Geller, Daniel S., and J. David Singer. *Nations at War: A Scientific Study of International Conflict*. Cambridge, UK: Cambridge University Press, 1998.

Gilli, Andrea, and Mauro Gilli. "The Diffusion of Drone Warfare? Industrial, Organizational, and Infrastructural Constraints." *Security Studies* 25, no. 1 (2016): 50–84. doi:10.1080/09636412.2016.1134189.

Gilpin, Robert. *War and Change in World Politics*. Cambridge, UK: Cambridge University Press, 1981.

Gray, Colin S. *Strategy for Chaos: Revolutions in Military Affairs and the Evidence of History*. London: Frank Cass, 2002.

Grissom, Adam. "The Future of Military Innovation Studies." *Journal of Strategic Studies* 29, no. 5 (2006): 905–934. doi:10.1080/01402390600901067.

Hall, Abigail R. "Drones: Public Interest, Public Choice, and the Expansion of Unmanned Aerial Vehicles." *Peace Economics, Peace Science and Public Policy* 21, no. 2 (2015): 273–300. doi:10.1515/peps-2014-0043.

Harmon, S.Y. "The Ground Surveillance Robot (GSR): An Autonomous Vehicle Designed to Transit Unknown Terrain." *IEEE Journal on Robotics and Automation* 3, no. 3 (1987): 266–279. doi:10.1109/JRA.1987.1087091.

Hern, Alex. 2015. "US Marines Reject BigDog Robotic Packhorse because It's Too Noisy." *Guardian*, December 30. https://www.theguardian.com/technology/2015/dec/30/us-marines-reject-bigdog-robot-boston-dynamics-ls3-too-noisy.

Hone, Thomas C., Norman Friedman, and Mark D. Mandeles. *American and British Aircraft Carrier Development, 1919-1941*. Annapolis, MD: Naval Institute Press, 1999.

Horowitz, Michael C. *The Diffusion of Military Power: Causes and Consequences*. Princeton, NJ: Princeton, 2010.

Horowitz, Michael C., Sarah E. Kreps, and Matthew Fuhrmann. "Separating Fact from Fiction in the Debate over Drone Proliferation." *International Security* 41, no. 2 (2016): 7–42. doi:10.1162/ISEC_a_00257.

Husseini, Talal. 2019. "In Pictures: The UK MoD's Future Robot Army." *Army Technology*, June 26. https://www.army-technology.com/features/future-military-robots-uk/.

Iagnemma, Karl, Shingo Shimoda, and Zvi Shiller. "Near-optimal Navigation of High Speed Mobile Robots on Uneven Terrain." *2008 IEEE/RSJ International Conference on Intelligent Robots and Systems*, 4098–4103, Nice, France, September 2008.

Judson, Jen. 2017. "Army Poised to Transform Ground Robotics Industry." *Defense News*, April 12. https://www.defensenews.com/smr/unmanned-unleashed/2017/04/12/army-poised-to-transform-ground-robotics-industry/.

Jun, S-P. "A Comparative Study of Hype Cycles among Actors within the Socio-Technical System: With A Focus on the Case Study of Hybrid Cars." *Technological Forecasting & Social Change* 79, no. 8 (2012): 1413–1430. doi:10.1016/j.techfore.2012.04.019.

Kenyon, Henry S. "U.S. Robots Surge onto the Battlefield." *Communications and Electronics Association* 62, no. 7 [Fairfax, VA, Signal, Armed Forces] (2008). https://www.afcea.org/content/us-robots-surge-battlefield

Kissinger, H A. "Arms Control, Inspection and Surprise Attack." *Foreign Affairs* 38, no. 4 (1960): 557–575. doi:10.2307/20029443.

Klafter, R.D. "Mobile Robots, Research and Development." In *International Encyclopedia of Robotics: Applications and Automation*, edited by Richard C. Dorf and Simon Y. Nof, 920–943. New York: Wiley and Sons, 1988.

Klare, Michael T. "Autonomous Weapons Systems and the Laws of War." *Arms Control Today* 49, no. 2 (2019): 6–12.

Krepinevich, Andrew. "Cavalry to Computer: The Pattern of Military Revolutions." *The National Interest* 37 (1994): 30–43.

Krieg, Andreas, and Jean-Marc Rickli. *Surrogate Warfare: The Transformation of War in the Twenty-First Century*. Georgetown, DC: Georgetown University Press, 2019.

Lamb, Gregory M. 2005. "Battle of the Bot: The Future of War?" *Christian Science Monitor*, July 27. https://www.csmonitor.com/2005/0127/p14s02-stct.html.

Lee, Jun Pyo. "Future Unmanned System Design for Reliable Military Operations." *International Journal of Control and Automation* 5, no. 3 (2012): 173–185.

Mahnken, Thomas G. *Technology and the American Way of War since 1945*. New York: Columbia University Press, 2008.

McNeill, William H. *The Pursuit of Power: Technology, Armed Force and Society since A.D. 1000*. Chicago: The University of Chicago Press, 1982.

Mearsheimer, John. *The Tragedy of Great Power Politics*. New York: W. W. Norton, 2014.

Michel, Arthur H., and Dan Gettinger. 2012. "Out of the Shadows: The Strange World of Ground Drones." *Blog post for the Center for the Study of the Drone at Bard College*, November 3. https://dronecenter.bard.edu/shadows-strange-world-ground-drones/.

Minkov, Yaniv, and Tal Oron-Gilad. "Remotely Operated Vehicles (Rovs) from the Bottomup Operational Perspective." In *Human-Robotic Interactions in Future Military Operations*, edited by Michael Barnes and Florian Jentsch, 211–228. Surrey, UK: Ashgate Publishing Limited, 2010.

Mizokami, Kyle. 2018. "Russia's Tank Drone Performed Poorly in Syria." *Popular Mechanics*, June 18. https://www.popularmechanics.com/military/weapons/a21602657/russias-tank-drone-performed-poorly-in-syria/.

Morris, Jefferson. "Military Projects 4,000 Robots in Theatre in FY'06." *Aerospace Daily and Defense Report* 217, no. 26 (2006): 4.

Murphy, R. R. *Introduction to AI Robotics*. Cambridge, MA: MIT Press, 2000.

Murray, Williamson. *Military Adaptation in War: With Fear of Change*. Alexandria, VA: Institute for Defense Analyses, 2009.

Nguyen-Huu, Phuoc-Nguyen, Joshua Titus, Dawn Tilbury, and G. Ulsoy. "Reliability and Failure in Unmanned Ground Vehicle (UGV)." *GRRC Technical Report* (2009).

Novichkov, Nikolai. "New Russian Combat UGV Breaks Cover, Uran Readies for Service." *IHS Jane's International Defense Review* 9, 2016.

Peoples, Columba. "Sputnik and 'Skill Thinking' Revisited: Technological Determinism in American Responses to the Soviet Missile Threat." *Cold War History* 8, no. 1 (2008): 55–75. doi:10.1080/14682740701791334.

Posen, Barry. *The Sources of Military Doctrine: France, Britain, and Germany between the World Wars*. Ithaca, NY: Cornell University Press, 1984.

Powell, Robert. *In the Shadow of Power: States and Strategies in International Politics*. Princeton, NJ: Princeton, 1999.

Rossiter, Ash. "Drone Usage by Militant Groups: Exploring Variation in Adoption." *Defense & Security Analysis* 34, no. 2 (2018): 113–126. doi:10.1080/14751798.2018.1478183.

Rossiter, Ash. "High-energy Laser Weapons: Overpromising Readiness.." *Parameters* 48, no. 4 Winter (2018–2019): 33–44.

Rossiter, Ash. "Participation in Warfare: Are We Witnessing Paradigmatic Change?" *Journal of Military and Strategic Studies* 19, no. 3 (2019): 103–123.

Rossiter, Ash. "Robotics, Autonomous Systems, and Warfare." *Small Wars & Insurgencies* 31 (2020): 4.

Ryan, Mike. *Human-Machine Teaming for Future Ground Forces*. Washington D.C: Center for Strategic and Budgetary Assessments, 2018.

Scharre, Paul. *Army of None: Autonomous Weapons and the Future of War*. New York: WW Norton & Company, 2018.

Seidel, R W. "From Glow to Flow: A History of Military Laser Research and Development." *Historical Studies in the Physical and Biological Sciences* 18, no. 1 (1987): 1. doi:10.2307/27757598.

Shaker, Steven M., and Alan R. Wise. *War without Men: Robots on the Future Battlefield.* Dulles, VA: Brassey's, 1988.

Shaker, Steven M., and Alan R. Wise. *War without Men: Robots on the Future Battlefield.* Dulles, VA: Brassey's, 1988.

Sharkey, Noel. "Cassandra or False Prophet of Doom: AI Robots and War." *IEEE Intelligent Systems* 23, no. 4, July/August (2008): 14–17. doi:10.1109/MIS.2008.60.

Singer, Peter W. *Wired for War: The Robotics Revolution and Conflict in the 21st Century.* New York: Penguin Books, 2009.

Sloan, Elinor. "Robotics at War." *Survival* 57, no. 5 (2015): 107–120. doi:10.1080/00396338.2015.1090133.

Smit, W. A. "Science, Technology and the Military: Relations in Transition." In *Handbook of Science and Technology Studies*, edited by S. Jasanoff, G. E. Markle, J. C. Peterson and T. Pinch, 598–626. London: Sage, 1995.

Smuda, B., E. Schoenherr, H. Andrusz, and G. Gerhart. "Deploying of Omni-directional Inspection System in Iraq and Afghanistan." *US Army TARDEC*, 2004.

Tiron, Roxana. 2002. "Ground Robotics Experience Bumpy Ride." *National Defense*, September 1. https://www.nationaldefensemagazine.org/articles/2002/9/1/2002september-ground-robots-experience-bumpy-ride

Tsitsimpelis, Ioannis C., C J. Taylor, B. Barry Lennox, and M J. Joyce. "A Review of Ground-based Robotic Systems for the Characterization of Nuclear Environments." *Progress in Nuclear Energy* 111 (2019): 109–124. doi:10.1016/j.pnucene.2018.10.023.

Van Creveld, Martin. *The Transformation of War.* New York: Free Press, 1991.

Van Lente, Harro, Charlotte Spitters, and Alexander Peine. "Comparing Technological Hype Cycles: Towards a Theory." *Technological Forecasting and Social Change* 80, no. 8 (2013): 1615–1628. doi:10.1016/j.techfore.2012.12.004.

Von Hippel, Eric A. "'Sticky Information' and the Locus of Problem Solving: Implications for Innovation." *Management Science* 40, no. 4 (1994): 429–439. doi:10.1287/mnsc.40.4.429.

Wong, Zhuhai K. 2018. "China's Norinco Rolls Out New Combat, Reconnaissance, and Logistics UGVs." *Jane's 360*, November 30. https://www.janes.com/article/84935/china-s-norinco-rolls-out-new-combat-reconnaissance-and-logistics-ugvs.

York, Herbert F. *The Advisers: Oppenheimer, Teller, and the Superbomb.* San Francisco, CA: W.H. Freeman and Company, 1976.

Young, S. "Military Explores Robotic Technology to Keep Soldiers from Harm's Way." *Unmanned Systems* 8, no. 4 (1990): 39–42.

Zhang, G., C. Duncan, J. Kanno, and R. Selmic. "Unmanned Ground Vehicle Navigation in Coordinate-Free and Localization-Free Wireless Sensor and Acutuator Networks." *Proceedings of the IEEE International Conference on Control Applications*, Yokohama, Japan, September 2010.

Artificial intelligence, big data and autonomous systems along the belt and road: towards private security companies with Chinese characteristics?

Peter Layton

ABSTRACT
China is pressing ahead with ambitious plans to create a massive infrastructure network connecting it with many countries across the globe. Some of the 'belt and road' infrastructure will however, run through regions convulsed by chronic civil unrest, substantial criminality and incipient insurgencies and need protection by China's private security companies. Simultaneously the Chinese state is undertaking a major 'anti-secession and counter-terrorism' campaign in Xinjiang using a variety of high-technology means: artificial intelligence, big data, wireless connectivity, autonomous systems and robotics. The demand and supply sides seem to be in sync, suggesting Chinese private security companies will soon use a suite of advanced information technology systems with a proven employment doctrine across much of Central Asia, South Asia and Africa. Such a future may be plausible but it is by no means certain as various factors may yet thwart China's private security companies.

Introduction

The tasks modern private security companies (PSCs) undertake are once again changing. In the 2000s, driven by the demands of America's Middle Eastern wars, PSCs went into conflict zones to actively support national armed forces, government agencies and aid organisations.[1] In the 2010s, driven by the demands of Russian national prestige, PSCs became instruments of the state, directly advancing national interests through irregular warfare operations.[2] In the 2020s, driven by the demands of China's geo-economic Belt and Road Initiative (BRI), PSCs seem set to advance into a new field: the long-term protection of major commercial investments across regions beset by chronic civil unrest, substantial criminality and incipient insurgencies. This new direction appears likely to draw heavily on information technology advances including artificial intelligence, big data, cloud computing, wireless connectivity, autonomous systems and robotics.

The term 'private security companies' is a broad one that encompasses small, local firms operating domestically to large international companies offering a range of offshore services including the ability to operate in conflict zones and use armed force. This article focuses towards the latter end of the continuum, albeit in many cases such companies developed out of those small local firms. The term 'private military companies' is sometimes used for this harder end of the PSC continuum but in the Chinese case 'private security companies' is more common. It is also arguably more accurate given that the BRI spans several dozen nations, each with their own unique security circumstances. In this, the broadening of conceptions of what the military domain encompasses plays a part. Military matters now incorporate many operations other than war, including in the cyber realm. The differences between peacetime and wartime activities have blurred as the notion of omnipresent security threats and challenges has gained traction.

In this, PSCs are not mercenaries. Structurally, PSCs are normal corporate entities operating in the global marketplace, endeavouring to be profitable and grow their businesses. They offer a range of services, have strategic plans and seek long-term relationships with clients. In contrast, mercenaries focus on the short-term profits individuals could make through threatening or applying violence. Individual not group competence is prized and the services offered are very limited. If mercenaries privatize conflict, PSCs corporatize it.

In the new direction PSCs appear to be evolving towards, the lead role is being played by Chinese PSCs. In turn the development of these PSCs is being driven by China's BRI. Announced in 2013, the state-led BRI is a one trillion-dollar global connectivity project encompassing infrastructure, energy, finance and trade. It is a vast mercantilist endeavour that aims to grow the Chinese economy through facilitating extensive international trade across Eurasia and Africa. It has grown into being of a size that it now shapes how China engages its near region and the wider world. By mid-2019, 126 countries and 29 international organisations had signed cooperation agreements, more than 3,000 projects were listed as within the BRI umbrella and some 60 percent of these were underway.[3] For Chinese President Xi Jinping, the BRI is 'the project of the century.'[4]

Some BRI countries are not uncomplicated investment destinations, having social conflicts, civil unrest and incipient insurgencies. As more Chinese companies invest in these weak states, security concerns are growing. These worries are being reinforced by an increasing number of attacks on Chinese worksites and workers.[5]

Moreover, there is evidence that BRI itself can worsen security in the countries and regions it crosses. The cost of BRI infrastructure projects can impose substantial debt burdens on host countries, while creating opportunities for graft, corruption and elite rent-seeking behaviour. Such actions can deepen existing tensions, creating social instability and anti-Chinese sentiment, as has happened in Sri Lanka and Pakistan.[6] While China would prefer that local governments manage security completely, not all are sufficiently capable and some need Chinese assistance.

An option is to deploy People's Liberation Army (PLA) elements but this is unappealing. Such a step would create the perception internationally that the BRI was a geo-strategic initiative motivated by great power politics not an economic one aiming to raise everyone's prosperity as China asserts. Moreover, deploying the PLA would mean China abandoning its longstanding policy of military non-interference in other countries.[7] Many would perceive China as an imperialist power not a trading partner. PSCs appear the answer to this dilemma. Chinese analysts however, consider the ability of current Chinese PSCs to undertake such international operations to be poor, with the sector still immature.[8]

This article examines the way Chinese PSCs might develop to meet the BRI's demands, in particular through employing advanced technology such as large-scale urban surveillance networks, big data, artificial intelligence, biometrics, drones, robotics and remote support. The article initially assesses the current characteristics of the Chinese PSC industry sector with its well-funded state-owned enterprise customer base, a reliance on ex-military personnel, a preference to stay within the company compounds the PSC is securing and a symbiotic relationship with the state. The second section discusses how advanced technology is being applied to digitise and automate security in China's Xinjiang province, Pakistan's 'safe-city' program and Zimbabwe's national facial recognition program. This technology has relevance to the BRI's security problems and prudently applied might address some current limitations of Chinese PSCs.

The third section combines the understanding of the Chinese PSC sector with the contemporary Chinese security concepts and technology to suggest a new Chinese PSC model and the capabilities it might possess. This new model is likely to build on, and augment, the currently employed security technology but is not without problems and shortcomings as the fourth section discusses. Concerns include: that a one-size fits all approach may be inappropriate given the diversity of BRI countries; that BRI security is not just a matter restricted to the specific BRI infrastructure sites themselves; that the Chinese strategy in calming Xinjiang province is more than simply applying security surveillance technology and is arguably not replicable elsewhere; and that the Chinese Communist Party (CCP) has worries about PSCs becoming well-trained, independent, militia-like groups.

The conclusion brings the discussion together to decide that Chinese PSCs embracing advanced security technologies to meet the BRI's demands appears plausible. A new form of PSC appears likely to emerge that is different to the American or Russian models. China's new digitalised and automated PSCs may become 'PSCs with Chinese characteristics.'

Importantly, for ease of discussion in this article the terms 'government' and 'state' are used in the conventional way that sees a distinction between society, the state and the government. However, China is not a conventional country. Instead the CCP rules in a uniquely idiosyncratic manner well captured by its vision of building 'socialism with Chinese characteristics.' When encountering the terms 'state' or 'government' this refers to the Party and its manifestations.

Chinese PSC industry today

In China, the market for domestic security services is saturated making the new opportunities offered by the BRI attractive. Going international though, is a complex undertaking requiring an adequate sized company and appropriate capabilities. In 2016, only 20 Chinese PSCs were offering international services. At that time these companies had about 3200 employees offshore; by comparison only 2600 PLA personnel were then deployed offshore on UN peacekeeping missions.[9]

The potential international market is very large with almost one million Chinese workers offshore, most employed on large official projects.[10] Unsurprisingly for some Chinese PSCs, the BRI is the basis of their entire market strategy.[11] John Jiang, managing director of Chinese Overseas Security Group, a small consortium operating in six countries, notes that: 'One Belt, One Road is a lifetime of work for us, in eight years' time [2025], we want to run a business that can cover 50 to 60 countries, which fits with the 'One Belt, One Road' coverage ....'[12]

Offshore security is a costly business with the principal customers for Chinese PSCs large State Owned Enterprises (SOE). As an example, the China National Petroleum Company earmarks 3–5 percent of its budget for investments in risky countries for security-related activities. In places where security conditions are extremely unsafe, such as in Iraq, budget allocations can approach 20 percent. In 2015, the company spent more than 2 USD billion (USD) on security. In contrast, smaller private, businesses often underestimate the risks abroad, favour minimising costs and typically allocate only 1 percent of their budget to security.[13]

In the field, Chinese PSCs generally adopt a similar operating model. As they often lack the necessary language skills, intelligence gathering capabilities and connections to local governments, most only operate within fenced compounds. The compound is then defended from external threats by local PSCs, militias or non-Chinese PSCs.[14] In Pakistan for example, Chinese personnel act as security managers inside the Chinese workers' gated compounds, while the Pakistani army and foreign PSCs guard the outside.[15]

For Chinese SOEs, this model has some advantages in that security is managed by Chinese PSCs rather than by the SOE directly hiring local or foreign PSCs. This outsources any language or cultural issues to the PSC. Moreover, SOEs have more trust in Chinese PSCs being within the company compound as there is a strong belief that they will protect an SOE's secrets better than the other PSC options. In Africa, SOEs further worry that local PSCs, while lower cost, could become part of the security problem and not the solution, and may become entangled in tribal feuds or local political disputes.[16]

Chinese PSCs often differentiate themselves from their foreign counterparts by claiming their employees are unarmed but this is a more complicated matter than it may appear. Within China, there are laws that regulate the arming of domestic PSCs but these do not apply offshore where local laws apply. In Iraq for example, the Chinese PSC staff are not allowed to carry guns, albeit workarounds have been devised. Tao Dexi, a contractor with Dingtai Anyuan International Security, notes that: 'Chinese security companies have always carried out security missions via local teams … .[however] under extreme emergencies, Chinese security staff can borrow guns from the local security staff.'[17]

The Chinese government now discourages the use of guns by PSCs operating offshore, fretting that such actions could cause a political backlash in the host country against China. The Chinese government is keenly aware of the negative reputational effects that the Blackwater PSC scandal created for the U.S. government in the 2007 Nisour Square shooting in Iraq and wishes to avoid such notoriety.[18] Already, several incidents in Africa and Central Asia have led to Chinese managers firing locally acquired weapons at groups of host nation workers.[19]

The Chinese state has a strong influence on its nation's PSC development. With the PSC's principal customers being SOEs, the state can steer the SOEs to award security contracts as deemed appropriate. At the moment, the state is keen to see Chinese PSCs rather than foreign PSCs gain SOE contracts. HXZA, for example, has gained a near monopoly on security for Cosco Holding and China Shipping Container Lines, China's two largest SOE shipping groups. A foreign PSC observed that 'they clearly have very solid relations to the state, considering how loyal their customer base is. And they are not that cheap.'[20]

There is a strong business referral network in which SOEs and former Chinese police and PLA officials share long-term friendships, common interests and CCP membership. The contracting procedures and the security tender awards can then come to rely more on the personal networks of a PSC's CEO than on their company's capabilities.[21] The PSCs themselves almost exclusively recruit PLA veterans.[22]

Given these close relationships, Chinese PSCs do not see themselves being hired by foreign clients; they are instead Chinese firms for Chinese customers. In this there is a reciprocal assumption that agreements the Chinese state has with host country governments can be used to shield Chinese PSC from trouble if needs be.[23]

This strong state-PSC relationship is not inviolate. While Chinese PSCs use ex-PLA staff, these individuals are generally new to the PSC industry and relatively inexperienced in managing conflict situations in unstable countries like Sudan and South Sudan.[24] In 2014, a Hong-Kong based, Chinese-funded, American-operated private security group titled Frontier Services Group (FSG) emerged and quickly gained numerous BRI project SOE clients. FSG was initially headed by Erik Prince, founder of the Blackwater PSC but is now chaired by Chang Zhenming of CITIC, its biggest investor and an SOE. Prince remains with the company as Deputy Chairman.[25] The Chinese state appears interested in pushing its PSCs to become more capable and more competitive by accessing foreign expertise.

The differences between Chinese PSCs and a Western PSC like Blackwater are stark and suggest how much Chinese firms need to mature. Tian Buchou, with almost two decades managing security for Chinese companies in the Middle East and Africa, observed after operating with Blackwater that:

> Unlike the Chinese security teams who … are like doorkeepers, the multinational and multilingual Blackwater guards were well trained and provided personnel and cargo protection services … [Blackwater] had a comprehensive operational system covering logistics, weapons, high-technology and even medical support. … More than 80 per cent of Chinese security personnel have just a basic education … and are directly led by people who are just military enthusiasts and love playing war games.[26]

Automating security

In far western China, Xinjiang province, officially titled the Xinjiang Uygur Autonomous Region, is both a BRI cornerstone where three major transit routes intersect and a major centre for trialling new security technologies. These commercial information technology innovations are already spreading along the BRI, particularly into Pakistan and parts of Africa. Security is being digitised and automated through large-scale urban surveillance networks, big data, artificial intelligence, facial recognition, biometrics, ubiquitous GPS tracking and smartphone spyware.

In terms of technology, the core of the large ecosystem of social monitoring and control in Xinjiang is the Integrated Joint Operations Platform (IJOP). It is an optimised security variant of a military information network system of systems developed for the PLA by a subsidiary of China Electronics Technology Group Corporation (CETC), a major state-owned military contractor.[27] Like the military systems, IJOP electronically gathers, stores and shares digital information.

IJOP integrates data collected from numerous widely deployed human sources and machine sensors to create a near real-time common operating picture that guides security activities.[28] The data continually collected by IJOP is further used by being structured into a machine readable database format and then stored to become so-called 'big data,' a very large data set able to be analysed computationally to uncover patterns, trends and relationships. The big data is then analysed using artificial intelligence algorithms so that IJOP can provide automated identification of individual people of interest detected by the machine sensors, and a prediction of their possible near-term activities. As more and more data is collected, the algorithms are constantly refined and their accuracy and reliability improved through using machine-learning techniques. While data is collected on everyone in Xinjiang, the database is apparently optimised for the particular groups and individuals of interest.

The machine sensors that collect the data automatically feed into IJOP are diverse and spread across the physical and digital worlds. Daylight and infrared facial recognition capable CCTVs are dispersed across the urban environment; 'wifi sniffers' collect the unique identifying addresses of computers, smartphones, and other networked devices; visitor management systems at numerous checkpoints monitor ingoings and outgoings; biometric security systems check individuals, making use of the DNA samples, fingerprints, iris scans, voice prints and blood types collected from all Xinjiang's citizens aged between 12–65; cyber systems check the use of Virtual Private Networks and encrypted communication tools such as WhatsApp, Viber and Telegram; speech recognition software screens landline and mobile telephone calls; walking gait scanners supplement facial recognition systems; the obligatory Clean Net Guard app on the Uyghur ethnic groups' smartphones monitors what is said, read, and written and who they connect with; commercial and 'spybird' drones provide video aerial surveillance; GPS trackers installed in private vehicles track movements; and individuals' electricity and petrol usage is scrutinized.[29]

Using this sensor array, the IJOP autonomously tracks the path of individuals across both space and time, looking for changes to the normal pattern of life. When an irregularity or deviation from what IJOP has earlier collected and stored in its database occurs, the system flags this micro-clue to the security authorities as suspicious, prompting an investigation. These deviations can include individuals contacting persons of concern, whether locally or internationally, or material changes like activating a new mobile phone. IJOP effectively erects a series of virtual fences that limits system-specified individuals' freedom of movement.[30]

The security services in the field interact with IJOP through a CTEC-developed app on their Android smartphones. The app collects and sends back to the IJOP Face++ facial recognition information and local area wireless network data. The app is also used in reverse by the IJOP to assign tasks to the security personnel, monitor their performance and score their success in fulfilling tasks. IJOP aims to not just guide the security personnel but also control them ensuring duties are effectively and efficiently carried out.[31]

The IJOP is always running in the background across Xinjiang, always learning more through using artificial intelligence, big data and cloud technology. The overall intent is both to make security more effective but also more efficient to reduce operating and personnel costs. In this, commercial companies are involved not just in the development but also the operation of the IJOP and its associated security technologies.

A tender released in mid-2017 for a 'safe city data centre' in Shache County, Xinjiang is an example of this. The centre sought was to have five floors, four of them underground, and to include IJOP, a command centre, large video displays, videoconference facilities, wireless networked control systems, computer rooms, cloud storage, and high quality security access measures. The centre was to be acquired through a public-private partnership using the 'Build-Operate-Transfer model' operating over a ten-year period. The first year was for construction, the remaining nine involved the winning bidder operating and maintaining the centre, with the centre handed over to the local authorities on contract completion.[32]

The Xinjiang security technologies are now being exported commercially. Perhaps unsurprisingly, the first exports are across the border into what Chinese Premier Li Keqiang considers the BRI's flagship project: the 63 USD billion China-Pakistan Economic Corridor (CPEC).[33] Pakistan is no easy country for the BRI to pass through, given its chronic civil unrest and endemic societal turbulence. China's leaders are though publicly committed to the BRI and have spent considerable political capital advocating for it. China is to a certain degree trapped given Pakistan's centrality to the project, making security within the CPEC crucial.

China is supporting numerous so-called 'safe city' projects to improve security within Pakistan's major urban centres by providing soft, low interest rate loans and urging Chinese SOEs to become involved. Safe city projects effectively install much of the Xinjiang digital security system with its dispersed machine sensors across Pakistani cities, including Islamabad, Peshawar, Lahore, Quetta, Karachi, and Gwadar.

Huawei has been involved in building the safe city system in Lahore. This has involved establishing a large, 7000 sqm Safe-City Integrated Command & Communication Centre that uses an IJOP-like system incorporating artificial intelligence, big data and cloud technology. The Lahore safe city system uses some 8000 high-grade CCTV cameras, 4G wireless connectivity, facial recognition, automated vehicle number plate recognition, multiple tracking options, integrated communication platforms, geographic information systems and specialised apps on security personnel's smartphones.[34]

Along with digitising Pakistani city security, Chinese SOEs are deeply involved in improving digital connectivity between the two countries, primarily through a cross-border fibre-optic cable link. Internet traffic is routed through China with discussion now focussed on adopting Chinese internet regulations and installing a Chinese-style national internet firewall.[35] Such connectivity effectively makes Pakistan an outlier of the Chinese network.

In Africa, the technologies pioneered in Xinjiang are also finding customers. In 2018, CloudWalk, a Guangzhou-based tech startup that has received considerable state finance, agreed a strategic cooperation framework agreement with Zimbabwe to build a national facial recognition program. The agreement as part of the BRI will see the technology primarily used in security and law enforcement.[36] Movement through Zimbabwe's airports, railways, and bus stations will be monitored using a facial recognition database integrated with biometrics.[37]

The CloudWalk agreement is the first Chinese artificial intelligence project in Africa and, as such, the technology will need some modification for local conditions.[38] The facial recognition database used in Xinjiang is extensive but inappropriate for an African population. Machine learning using a large local database is essential to optimise the Chinese artificial intelligence software for the Zimbabwean urban environment. Accordingly, as part of the agreement Zimbabwe will send the facial data on millions of its citizens as captured by CCTV cameras to CloudWalk so it can improve the artificial intelligence systems ability to distinguish between dark-skinned individuals.[39]

In this, the Zimbabwean database is not the only machine learning option for China's artificial intelligence systems engineers. In 2017, Chinese company Transsion became the dominant company in the African mobile smartphone market, mainly through optimizing its technology for the African consumer. The smartphones support multiple African languages, dual SIM functionality and cameras that calibrate their exposure settings for darker skin tones, allowing face details to be better captured. This latter feature is important for the company's new Tecno Camon X Pro that includes facial recognition technology. Transsion will now collect data on millions of customers, allowing the company to improve its facial recognition capabilities through machine learning.[40]

The need to collect African-specific images to improve the robustness and reliability of 'big data' facial recognition systems is readily apparent. The same issue will doubtless exist in Pakistan where the large scale collection of images by CCTV systems collated with biometrics – fundamental to the safe cities' projects – will be inevitably feed back into Chinese artificial intelligence machine learning software to improve its performance. The broadband fibre-optic link seamlessly connecting Pakistan and Chinese digital networks should allow such artificial intelligence training to be undertaken easily and quickly.

A new model Chinese PSC?

The Chinese government wants Chinese PSCs to expand to meet BRI demand, but such growth is problematic given staffing constraints. There are limited numbers of personnel with suitable expertise available; few have more than five years' experience in the field and most have only worked in a permissive security environment.[41] The impact of the skill shortage can be observed in Frontier Services Group's assessment of much current Chinese PSC planning for higher risk areas. This often neglects 'risk assessments, training, emergency action plans, medical evacuation, tracking, communication, process and procedures, security design and hardware ....'[42] On the other hand, the Chinese state and SOEs have an interest in making Chinese PSCs succeed. Generous funding may be available with the right business case.

Improvements have been forced upon Chinese PSCs providing security for energy industry SOEs operating in southern Iraq. The difficult security environment combined with the Iraqi government's requirements for licences and regulatory compliance have led to Chinese PSCs developing higher professional standards. Company capabilities and risk management have been enhanced and new technology introduced including unmanned aerial vehicles and tethered blimps.

This so-called 'Iraq Model' appears an important benchmark for the entire Chinese PSC industry and may be repeated along the BRI. An important aspect of this model transferable to security activities in other BRI countries is the specific objectives the PSCs involved in outhern Iraq have. These are straightforward and uncomplicated involving focussing on safeguarding personnel and guarding infrastructure. The broad intent is simply to protect SOE infrastructure and, in the event of a crisis, evacuate Chinese personnel.[43]

The various factors of increasing demand for security, skilled personnel shortages, clearly defined security objectives, and the usual commercial drive for improved productivity and profitability all combine to suggest that Chinese PSCs will aim to exploit advanced security technology. The digitisation and automation of security in Xinjiang offers a starting point to consider how Chinese PSCs operating along the BRI might adopt new technology.

The Xinjiang suite of security technologies is comprehensive, spanning an integrated central command and control platform (IJOP), urban surveillance networks, big data, artificial intelligence, voice and facial recognition, biometrics, ubiquitous GPS tracking and smartphone spyware. In broad terms, the suite is best suited for a specific location where the various systems can be established and optimised for the local situation. In the BRI though, this may be more difficult than it seems.

Across the BRI there will be some sites that will be discrete and remote, but many others will be within urban areas and some will be very large facilities like ports and airfields. Moreover, the BRI's connectivity is reliant on developing and keeping secure nation-wide transportation systems. For example, CPEC is a network of roads, rail, pipelines and fibre-optic cables that crosses Pakistan from the Chinese border to the Indian Ocean, near the Iranian border.

The BRI brings a new emphasis on securing areas rather than fenced company compounds. Such a shift may take some time to be fully understood. Frontier Services Group noted that many Chinese SOE customers today have misconceptions about operating in high-risk countries:

> It is not as simple as hiring a few guards [...] With many of the [belt and road] projects covering thousands of kilometres of road, rail, pipeline or hectares of oil field, building a wall or fence is simply not an option.[44]

Guarding areas is a difficult task, as it traditionally requires large numbers of people; a luxury Chinese PSCs operating offshore do not have. The BRI requirement for securing areas may not be feasible without employing Xinjiang suite technologies as a baseline.

The suite can be expected to have new technologies grafted-on as they become available. A number of changes and new additions appear useful for Chinese PSCs securing the BRI.

The easiest is simply increasing the numbers of machine sensors in the surveillance network. The Xinjiang system pioneered the use of large numbers of CCTVs capable of facial recognition and in so doing reduced costs through mass production. Costs continue to plummet allowing dramatic increases in the size of CCTV sensor fields. Singapore now plans to install facial recognition cameras on the 110,000 lampposts across the country and is running trials, with three major Chinese firms interested in bidding.[45] PSCs at BRI facilities can increasingly be relatively profligate in their use of CCTVs and, given low replacement costs, much less concerned about CCTV accidental losses and deliberate damage.

CCTV facial recognition requires passing the collected images to a central command and control platform where an artificial intelligence algorithm compares these against an optimised database. For a PSC task, such a database is easier to develop, as it only needs to ensure authorised people are within specific areas rather than trying to identify a single person in large, crowded urban areas. A comprehensive biometric database can be progressively created as each person is approved and authorised to have access. Over time the database will become more extensive and also more accurate, with fewer false positives as machine learning occurs.

Further downstream as 5G networks become available they will offer the potential to deploy hundreds, perhaps thousands, of Internet of Things (IoT) sensors across fixed sites. Such sensors might be quite simple in surveilling only a small area within a narrow electromagnetic band, but quantity can make up for quality shortcomings. The future Chinese PSC is likely to deploy a cloud of IoT sensors around a BRI location, all sending information back wirelessly to the central command and control platform for artificial intelligence algorithms to process, filter, fuse and display as a near real-time picture of the local area.

Such sensor fields can also send information back from authorised personnel to the command and control platform, allowing the personnel to be monitored and tracked in real time. Fitbit activity trackers and certain smartphones apps offer such capabilities now that are both robust and low cost.

The Xinjiang system now connects personnel in the field to the IJOP central command and control platform through smartphone apps. Such human-machine teaming is being taken further with the introduction of wearable technology, in particular augmented reality glasses. When wearing such equipment, PSC staff could access in near real-time facial identification and vehicle plate information held within the company's cloud-based database. Augmented reality glasses have been recently introduced in Xinjiang province.[46]

The most significant technical enhancement to the overall Xinjiang suite is likely to be further applications of artificial intelligence. In general terms, systems like the IJOP are excellent at gathering, storing and sharing information but less successful at processing and contextualising this information. The networks can be somewhat overwhelmed by the volume, velocity and variety of data being collected, affecting their ability to turn the data quickly into useful intelligence.[47] This shortcoming is partially overcome in Xinjiang by the simple expedient of employing large numbers of staff but this is not a practical solution for Chinese PSCs operating across the BRI. Incorporating artificial algorithms in IJOP or equivalent systems can help address these issues through:

- improving the integration and processing of information;
- adding new data fusion techniques to improve situational awareness;
- applying machine learning to remote sensing to enhance change detection;
- using 'expert systems' for high-level decision support;
- using machine learning for predictive maintenance and fault forecasting; and
- using virtual and augmented reality for simulations and training.[48]

Developments beyond the Xinjiang suite are likely to be in two main areas: mobile autonomous systems and adding remote support. The first will aim to allow PSCs to provide security over much larger areas, be able to quickly change the areas being surveilled and to be able to focus on a specific location in greater detail when necessary. Mobile systems may be essential when securing large ports and airfields and in guarding transportation and communication networks. The second area for development, remote support, will aim to improve the effectiveness and efficiency of the overall security system. Efficacy becomes significant when it is considered that the BRI is a long-duration project, potentially lasting generations.

A variety of mobile autonomous systems are becoming available and in this case may be categorised as drones, unmanned ground vehicles (UGVs) or robots. 'Drones' is a popular term for unmanned air vehicles. They can carry sensors able to feed information both to a local operator and the central command and control platform for analysis and display. The combination of low cost consumer electronic packages and plastic moulded air vehicles have allowed mass production and made small drones widely affordable. China is particularly active in hobbyist and commercial drone manufacture; the SZ DJI Technology Company (trading as DJI) has some 70 percent of the hobbyist drone global market share.[49]

For PSCs, commercial drones can now carry high-resolution camera systems able to identify individuals at ranges of about 800 metres from a drone flying at an altitude of 100 metres. While hobbyist drones have endurance constraints in being able to fly for only about 30 minutes, larger commercial drones can be connected to the ground via a power cable and have much longer flight time providing extended air surveillance. The connection to the ground can also include a high bandwidth communication cable.[50] Depending on requirements such a drone could be replaced by a tethered blimp carrying a similar sensor and communication package.

Free-flying drones can be flown forward to investigate emerging situations with some carrying audio systems to warn intruders. Chinese PSCs prefer not to use weapons and instead rely on local PSCs to use firearms however, armed drones may offer the alternative of letting robotics take over. ISIS deployed DJI hobbyist drones dropping hand grenades in the battle for Mosul.[51] These proved not just effective against small vehicles but also psychologically debilitating for those encountering them.[52] PSCs could do similarly; if not with explosives, then perhaps delivering non-lethal payloads such as smoke and teargas.

The tasks drones can cost-effectively accomplish is steadily increasing with growing interest recently in cargo drones. The overall impact is that Chinese PSCs at BRI sites could now deploy their own organic airpower that automates some tasks, reduces overall personnel requirements and keeps staff inside protected compounds to a much greater extent.

Drones have proliferated and become affordable but unmanned ground vehicles (UGV) are really only just emerging.[53] For some decades security companies have operated small bomb disposal robots but these have been highly specialised and in being tele-operated had a very short radius of action. Now such small robots are being fitted with wireless connections, diverse sensors and are all-terrain, including climbing stairs and exploring tunnels. Even so, with a range of about 500 m they have only limited utility for PSCs.

Israel's recent fielding of some 350 Segev UGVs however indicates what might be possible in the future. These UGVs, based on a Ford F-350 truck, can patrol autonomously along a designated off-road route and, if onboard sensors detect an intruder, inform a remote operator who can assess the images the vehicle sends. The Segev has a machine gun the remote operator can use to defend the vehicle or engage detected incursions. In this, Segevs are used where humans are not expected to be present rather than in crowded urban areas; Segevs are not self-driving cars but pre-programmed robots intended to detect changes on their patrol route.[54] Chinese PSCs might find such a vehicle useful, at least in its patrol mode, in being able to monitor large unpopulated areas with little human input. Automation can again reduce staffing requirements allowing fewer people to do more.

A variation on this approach is the use of robots for building or large facility security. An example is the Knightscope K-series robots: human-sized, bullet-shaped 140 kg Autonomous Data Machines that patrol programmed locations, have video and infra-red sensors and advise remote operators of any intruder detections.[55] While gradually gaining commercial acceptance, these robots face competition from the sharply declining cost of CCTVs. Large CCTV fields may produce results similar to indoor patrolling robots. The counter-argument is that CCTVs are fixed and may be able to be evaded; an autonomous mobile robot has a roaming physical presence that may deter intruders.[56]

The digitisation and automation of security that the Xinjiang suite, its enhancements, and mobile systems can provide indicates that PSCs could use technology to reduce the numbers of staff they deploy to trouble spots. Machines monitored and controlled remotely could take the place of personnel working in high-risk BRI areas. The staff structure profile of Chinese PSCs might change with more China-based staff, fewer personnel working in offshore locations and a large number of deployed robots. Such a remote manning construct further implies that the skills of the in-field personnel may change with fewer doing security tasks and more undertaking maintenance tasks, attending to the automated security system across the BRI site and beyond.

Implicit in remote manning is the ability of PSCs working at BRI sites to plug into broadband internet services and connect back to China. The investment in the Pakistan-China cross-border fibre-optic cable link fits this requirement well. This connectivity, in effectively making Pakistan part of the Chinese network, means PSC staff remaining in China can work electronically on digital security tasks in Pakistan almost seamlessly.

Reinforcing this change in PSC manning might be some new forms of support provided by the Chinese government. There is interest in establishing a state security agency to coordinate security for Chinese SOEs operating abroad. It is envisaged that the Ministry of State Security would collect intelligence and provide non-traditional security support to the new agency.[57] The current thinking sees the new agency coordinating Chinese PSC operations in BRI countries.

Such an agency would liaise with the host country government on the PSCs behalf together with sharing with the involved PSCs relevant Chinese national intelligence and data. The data might not just be tactical-level intelligence, but could include technical support for the digital and automated systems the PSCs are using including facial recognition databases and the latest software enhancements to computerised security systems. This directed state help would make PSC security activities in BRI countries more effective and more efficient.

Issues and shortcomings

A new model Chinese PSC would be more capable than the current approach albeit with some issues and shortcomings. The new model would be most appropriate for fixed sites occupied for an extended period where complicated digital systems could be installed and comprehensive databases built-up. In this, the new digitised and automated model is compatible with current Chinese PSCs that have a well-funded SOE customer base, predominantly employ ex-PLA personnel, prefer to stay within a fenced compound and be unarmed, and have a close relationship with the state. Moreover, the new model helps address the difficulty of expanding quickly enough to meet BRI demand by using the available personnel more efficiently and being able to access China-based support.

However, issues may arise in reducing interaction by the PSCs with local people in each BRI site. Automation in general, and the Xinjiang suite in particular, reduces people to objects to be managed. This approach may have shortcomings given the BRI is a generational project extending over decades. Forging local relationships might become more important over time with the current Chinese PSC approach of staying inside fenced compounds gradually becoming less efficacious.

Extending this line of argument highlights that in broad terms China is trying to apply the same PSC model across many different situations. BRI sites in some African states may suffer from criminality and disputes over poor pay for local employees; sites in Central Asia may experience clashes with the local populations over fears that an influx of Chinese workers might affect demographics; sites in South Asia may be attacked by insurgents wanting to embarrass the central government and gain political concessions.[58] Having a single dominant PSC approach might gain cost-efficiencies but may not be equally effective across countries. Alessandro Arduino is unsure most Chinese companies would realize this issue, noting that:

> With the exception of the Chinese oil and gas sector, there is a dangerous lack of interest among Chinese SOEs in understanding security. The fact that Chinese corporations still cannot differentiate anti-Chinese violence from political and criminal activity is an alarming example.[59]

Moreover, the new model Chinese PSC approach shares a significant shortcoming with the existing approach. While BRI sites might be well guarded, they remain dependent on the national security situation in the host country. Several BRI countries have governments that may be forcibly changed and the resultant civil strife could devastate BRI investments. Some 20 USDbn was invested in Chinese projects in Muammar Gadhafi's Libya; almost all of this was lost when the regime was overthrown.[60] The 2011 Arab Spring further revealed that paying money to a host country's armed forces to provide the outer protective screen of a fenced compound secured internally by Chinese PSCs does not work when the military disband or join a civil war.[61]

These examples may explain the Chinese state's current interest in helping existing authoritarian leaders install the Xinjiang suite in their countries to assist them staying in power. The suite is though only half of the Chinese strategy for countering social unrest. The Xinjiang suite has been developed to detect, locate and identify particular individuals allowing them to be detained. In Xinjiang province these individuals then enter a massive state re-education system while their family and local community undergo 'de-extremification.' The transformation of Xinjiang society is sought with the state purposefully acting to change the culture, customs, and thoughts of the local population.[62] The Xinjiang advanced technology suite alone is not enough.

The Xinjiang suite's technologies have been devised as a key element of the Chinese state's social control strategy but on the understanding that China has sizeable material means and massive human resources able to implement the strategy. As an example, in Xinjiang province the security measures taken include establishing some 7,500 'convenience police stations.' The province may be more highly policed per capita than East Germany was before its collapse in 1989.[63] It is a security model with Chinese characteristics that is unlikely to be easily replicated elsewhere.

Accordingly, the Chinese state has become more interested in mediating between parties involved in intra-state conflicts in South Asia, the Middle East and Africa. PSCs will not be enough to safeguard the BRI and helping autocrats stay in power may be problematic suggesting making mediation efforts is a worthwhile addition. The Chinese government is well aware that without adequate stability in key countries along the BRI, it may fail. Even so these efforts have attracted some criticism as being short-sighted. China's mediation efforts are mostly focusing on conflict management and stability preservation rather than long-term issue resolution.[64]

Through taking such seemingly pragmatic approaches to safeguarding its BRI interests, China is gradually moving away from its principle of non-interference in other state's domestic political structures. In its own way, deliberately seeding Chinese PSCs across the BRI is one step towards changing the Five Principles of Peaceful Coexistence that today constitute the pillars of China's foreign policy.[65]

In that regard, the Chinese state's development of PSCs may cause another even more fundamental problem. The CCP has historically worried about the private ownership of guns and other weapons – especially when they are in the hands of well-trained, tightly organized, militia-like groups.[66] Indeed, the PLA itself is not the Chinese state's army but rather the Party's. PSCs, as they expand and spread into dangerous countries, may end up armed, as they are now when protecting merchant ships against pirates. Some extend this, foreseeing the possibility that SOEs could use their large budgets to build-up specialized PSCs with military capabilities that might be useful in some future internal Chinese Communist Party power play.[67]

To counter such a move, in 2010 the Regulation on the Administration of Security and Guarding Services was introduced that stipulated the state must wholly own or be the controlling shareholder of any domestic armed PSC. In this way, Chinese PSCs permitted to carry weapons are administered and supervised by the state.[68] This future would seem to await those PSCs operating offshore as well. While the revolving door for personnel from the PLA into the PSC sector may presently be sufficient to guarantee their loyalty, as the PSCs grow the state is likely to move into some form of formal control.

Conclusion

The BRI demand for security services appears set to push Chinese PSCs towards adopting advanced security and social control technologies. This seems a plausible outcome given the various factors involved but is not certain.

In considering PSCs, Elke Krahmann usefully proposed two types of civil military relationships: republicanism and liberalism. Republicanism stressed the central role of the state. It was not only seen as more efficient and effective in providing security but also through centralisation could control security forces better. In contrast, liberalism stressed a small state sector that used market forces and private companies to both gain greater cost-efficiency and control security forces through fragmentation.[69] If republicanism saw the nation in arms, liberalism advocated contractorisation.

China is a country where the separation of the public and the private spheres implicit in Krahmann's model is at best immature.[70] Nevertheless, the Chinese PSC approach does tend towards liberalism and this will continue to deepen if the digitisation and automation argued in this paper is embraced. However, China has a choice.

The state could switch to the alternative republicanism approach. The state could take-over the PSC industry; this would be uncomplicated given the Chinese state often has large shareholdings in such companies. The PLA (or the People's Armed Police) could then take over running PSCs in BRI countries. If today's PSCs principal customers are SOEs, PSCs becoming SOEs themselves or even branches of the armed services does not seem implausible. Indeed, with the current Chinese Communist Party leaderships' attraction to statist policies and centralisation, bringing PSCs into the fold may at some stage appear attractive.

Against this, as noted earlier in this article, is that China could quickly become seen internationally as an imperialist power. Using PLA units rather than PSCs would generate significant pushback from BRI countries given many are acutely sensitive to any potential re-emergence of colonialism. There would be a real possibility that such a step could cause the BRI project to collapse. Even without this, such a move would inevitably increase bureaucratization and drive up service provision costs. Instead, for the foreseeable future, China seems intent on remaining with liberalism with its emphasis on, and enthusiasm for, defence contractorisation.

In this clash between a state-centred republican approach versus a commercial-centred liberalism approach, it is useful noting the direction Western PSCs have taken over the last decade. The key difference between Western and Chinese PSCs is as Blackwater owner Eric Prince observed in 2007 at the height of PSC boom during the Iraq War: 'We're a private company, and there's a key word there – private."[71] Chinese companies as noted are instead closely associated with the Chinese state; commercial factors are a secondary rather than the dominating concern.

Over the last decade, many major Western PSC companies have been bought out wholly or partly by private equity firms; these include Blackwater, Triple Canopy, DynCorp, Global Strategies Group and Gardaworld. Today, profitability takes precedence with the preferred strategies for achieving this being diversification into different industrial sectors for example cybersecurity, moving away from boots-on-the-ground to less personnel intensive services such as aviation, and creating new short-term companies for each specific contract.[72] The later means companies can 'rebrand in an instant.'[73] In analysing the British PSC industry, researchers noted: 'The industry is marked by a high turnover of companies; businesses are formed and then dissolved a short while later.'[74]

These various strategies means Western companies are less interested in pursuing costly investments in new high technology equipment. Their contracts are often only short term and the ability for the company concerned to disappear if incidents or financial issues require is a useful attribute.[75]

Moreover, Western companies can lower their labor costs by recruiting staff from lower salaried countries such as Nepal.[76] Chinese companies conversely prefer Chinese staff but, with these relatively scarce and expensive, investing in productivity-improving high technology to secure long-term contacts with Chinese SOEs makes financial sense. This predilection is reinforced by the inability of most Chinese PSCs to disappear given the Chinese state is a major shareholder.

The Western PSC business model has developed in a different direction to the Chinese PSC one. Moreover, the new Chinese digital technology relies to a great degree on 'big data' for its effectiveness; such large-scale data is owned by the state. Western PSCs simply do not have access to such information, especially as concerns the numerous countries they operate in. Chinese PSCs and the Chinese state have a symbiotic relationship quite unlike that Western PSCs have with their parent state in the post-Iraq War era.

The future seems promising for Chinese PSCs despite several matters of concern. In simple business terms, the principal unsettled issue seems to be gaining the funding necessary so the various companies can acquire advanced security and social control technologies. Given the Chinese state's current policy stance on PSCs and its understanding of their importance to the BRI such additional funding will doubtless be provided, most likely by the SOE customers.

Chinese PSCs can be expected to employ advanced security technologies as the companies grow in number and spread out across the BRI. In so doing, a new form of PSC will gradually emerge that is different to the American or Russian models. Chinese PSCs that are task-optimised to protect SOE commercial investments along the BRI will be a unique form. If the Chinese Communist Party seeks 'socialism with Chinese characteristics,' so China's new digitised and automated PSCs may become considered 'PSCs with Chinese characteristics.'

Notes

1. Singer, *Corporate Warriors*.
2. Marten, "Russia's use of semi-state security forces," 198–99.
3. Robinson, "Southeast Asia gains new leverage."
4. Zhou, "China's Belt and Road Forum."
5. Yang, "Securing China's Belt and Road Initiative," 2.
6. Legarda and Nouwens, "Guardians of the Belt and Road," 7.
7. Ramachandran, "Protecting BRI: China's Foreign Security Concerns."
8. Arduino, "China's Private Army," 41.
9. Ghiselli, "Market Opportunities and Political Responsibilities," 9.
10. Chan, "Why a private US military firm is of value to China's belt and road mission."
11. Legarda and Nouwens, "Guardians of the Belt and Road," 14.

12. Reuters, "Security firms to cash in protecting China's 'New Silk Road'."
13. Ghiselli, "Market Opportunities and Political Responsibilities," 8–9.
14. Arduino, "China's Private Army," 47–48.
15. Basit, "Terrorizing the CPEC," 708.
16. Arduino, "China's Private Army," 103.
17. Clover, "Chinese private security companies go global."
18. Yang, "Securing China's Belt and Road Initiative," 9–10.
19. Arduino, "China's Belt and Road Initiative," 18.
20. See note 17 above.
21. Arduino, "China's Private Army," 42.
22. See note 11 above.
23. Arduino, "China's Private Army," 47.
24. See note 11 above.
25. Adams, "Blackwater Founder Erik Prince's New Company Is Operating In Iraq."
26. See note 10 above.
27. "China: Big Data Fuels Crackdown in Minority Region."
28. Wang, "China's Algorithms of Repression."
29. Byler, "Ghost World,"; Wang, "China's Algorithms of Repression,"; and Chen, *China takes surveillance to new heights.*
30. See note 28 above.
31. See note 28 above.
32. Post translated by Google Translate: *1.65 billion | PPP Construction Project of Ping'an City, Shache County, Xinjiang.*
33. Boni, "Protecting the Belt and Road Initiative," 5.
34. Ibid., 7–8. Sponsored, "Here's How Punjab Police Integrated Command, Control and Communication Centre is Making Lahore Safer,"; PSCA, "Lahore to Get High Grade Surveillance Equipment Under Safe City Project,"; Huawei, "Huawei Announces Safe City Compact Solution to Protect Citizens in Small and Medium Cities."
35. Boni, "Protecting the Belt and Road Initiative," 9–11.
36. Chutel, "China is exporting facial recognition software to Africa."
37. Byler, "Ghost World."
38. Hawkins, "Beijing's Big Brother Tech Needs African Faces."
39. Pilling, "The fight to control Africa's digital revolution."
40. See note 36 above.
41. Arduino, "China's Belt and Road Initiative," 9.
42. See note 10 above.
43. Arduino, "Protecting Chinese oil."
44. See note 10 above.
45. Aravindan and Geddie, "Singapore to test facial recognition on lampposts."
46. Yang, "Chinese AR start-up develops smart glasses to help police catch suspects".
47. Layton, *Algorithm Warfare*, 37–38, 41–42.
48. This list draws on: Kania, *Chinese Military Innovation in Artificial Intelligence*, 10.
49. Thiercelin, "Beyond Personal Drones."
50.. Bogdanov, "Smart Face Control."
51. Rossiter, "Drone Drone Usage by Militant Groups," 118.
52. Layton, "Commercial drones."
53. Rossiter, "Bots on the Ground."
54. Layton, "Robot Wars," 14–15.

55. Joh, "Private Security Robots."
56. Ghaffary, "Is this robot really going to replace a security guard?"
57. See note 10 above.
58. Gafarov, "Rise of China's private armies," 43.
59. Arduino, "China's Belt and Road Initiative," 10.
60. Arduino, "China's Private Army," 72–73.
61. See note 10 above.
62. Doyon, "'Counter-Extremism' in Xinjiang."
63. See note 28 above.
64. Legarda and Hoffmann, "China as a conflict mediator."
65. Keast, "Speaker Interview."
66. See note 5 above.
67. Arduino, "China's Private Army."
68. Yang, "Securing China's Belt and Road Initiative," 9.
69. Krahmann, *States, Citizens and the Privatisation of Security*, 66, 72–76.
70. Ghiselli, "Market Opportunities and Political Responsibilities," 3.
71. *Blackwater*, 109.
72. Black, "What one company's vanishing act tells us about the private security industry."
73. Sean McFate on the US PSC industry quoted in Ibid.
74. Bruun, "Britain's private military and security industry examined."
75. Fielding-Smith et al., "A Security Company Cashed In on America's Wars."
76. Ibid.

Disclosure statement

No potential conflict of interest was reported by the author.

Bibliography

Adams, Rosalind. 2019. "Blackwater Founder Erik Prince's New Company is Operating in Iraq." *BuzzFeed News*. https://www.buzzfeednews.com/article/rosalindadams/blackwater-erik-prince-frontier-services-group-iraq

Aravindan, Aradhana, and John Geddie. 2018. "Singapore to Test Facial Recognition on Lampposts, Stoking Privacy Fears." *Reuters: World News.* https://www.reuters.com/article/us-singapore-surveillance/singapore-to-test-facial-recognition-on-lampposts-stoking-privacy-fears-idUSKBN1HK0RV

Arduino, Alessandro. 2017. "China's Belt and Road Initiative: The Evolution of Chinese Private Security Companies." *RSIS Working Paper Series.* Singapore: S. Rajaratnam School of International Studies.

Arduino, Alessandro. 2018. "China's Private Army: Protecting the New Silk Road." *The Diplomat*, 20 March. https://thediplomat.com/2018/03/chinas-private-army-protecting-the-new-silk-road/

Arduino, Alessandro. 2018. "Protecting Chinese Oil: A New Breed of Private Security Company Expands in Iraq." *The Arab Weekly*, August 5. https://thearabweekly.com/protecting-chinese-oil-new-breed-private-security-company-expands-iraq

"Autonomous Security." 2019. *Knightscope.* https://www.knightscope.com/

Basit, Saira H. "Terrorizing the CPEC: Managing Transnational Militancy in China–Pakistan Relations." *The Pacific Review* 32, no. 4 (2019): 694–724. doi:10.1080/09512748.2018.1516694.

Blackwater USA. "Hearing Before The Committee On Oversight And Government Reform House Of Representatives." *One Hundred Tenth Congress, First Session, October 2, 2007, Serial No. 110–89*, Washington: U.S. Government Printing Office, 2008.

Bogdanov, Vik. 2019 "Smart Face Control: How Facial Recognition Changes Different Industries." *Becoming Human: Artificial Intelligence Magazine.* https://becominghuman.ai/smart-face-control-how-facial-recognition-changes-different-industries-f03204496eaf

Boni, Filippo. "Protecting the Belt and Road Initiative: China's Cooperation with Pakistan to Secure CPEC." *Asia Policy* 14, no. 2 (2019): 5–12. doi:10.1353/asp.2019.0024.

Bruun, Laura, Iain Overton, and Elisa Benevilli. 2018. "Britain's Private Military and Security Industry Examined." *Action on Armed Violence*, December 20. https://aoav.org.uk/2018/britains-private-military-and-security-industry-examined/

Byler, Darren. 2018. "Ghost World." *Logic.* https://logicmag.io/07-ghost-world/

Chan, Minnie. 2018. "Why a Private US Military Firm Is of Value to China's Belt and Road Mission." *South China Morning Post*, July 15. https://www.scmp.com/news/china/diplomacy-defence/article/2155353/danger-zone-why-private-us-military-firm-value-chinas

"China: Big Data Fuels Crackdown in Minority Region." 2018. *Human Rights Watch*, February 26. https://www.hrw.org/news/2018/02/26/china-big-data-fuels-crackdown-minority-region

Chutel, Lynsey. 2018. "China Is Exporting Facial Recognition Software to Africa, Expanding Its Vast Database." *Quartz Africa.* https://qz.com/africa/1287675/china-is-exporting-facial-recognition-to-africa-ensuring-ai-dominance-through-diversity/

Clover, Charles. 2017. "Chinese Private Security Companies Go Global." *Financial Times*, February 27. https://www.ft.com/content/2a1ce1c8-fa7c-11e6-9516-2d969e0d3b65

Crofton, Black. "What One Company's Vanishing Act Tells Us about the Private Security Industry." *The Bureau of Investigative Journalism*, January 30 (2019). https://www.thebureauinvestigates.com/blog/2019-01-30/private-security-industry

Doyon, Jerome. 2019. "'Counter-extremism' in Xinjiang: Understanding China's Community-Focused Counter-Terrorism Tactics." *War On The Rocks*, January 14.

https://warontherocks.com/2019/01/counter-extremism-in-xinjiang-understanding-chinas-community-focused-counter-terrorism-tactics/

Fielding-Smith, Abigail, and Crofton Black, and The Bureau of Investigative Journalism. 2019. "A Security Company Cashed in on America's Wars—And Then Disappeared", *The Atlantic*, January 29. https://www.theatlantic.com/international/archive/2019/01/afghanistan-civilian-private-security/581263/

Gafarov, Odil. 2019. "Rise of China's Private Armies." *Chatham House: The World Today*, February & March. 41–43. https://www.chathamhouse.org/publications/twt/rise-china-s-private-armies

Ghaffary, Shirin. 2018. "Is This Robot Really Going to Replace a Security Guard?" *Vox*. https://www.vox.com/2018/10/8/17913420/security-robot-cobalt-robotics-knightscope-slack-yelp

Ghiselli, Andrea. "Market Opportunities and Political Responsibilities: The Difficult Development of Chinese Private Security Companies Abroad." *Armed Forces & Society* 20, no. 10 (2018): 1–21.

Hawkins, Amy. 2018. "Beijing's Big Brother Tech Needs African Faces." *FP: Foreign Policy*, July 24. https://foreignpolicy.com/2018/07/24/beijings-big-brother-tech-needs-african-faces/

Huawei. 2018. "Huawei Announces Safe City Compact Solution to Protect Citizens in Small and Medium Cities." *Huawei Technologies*, October 11. https://e.huawei.com/us/news/smart-cities/201810150942

Joh, Elizabeth E. "Private Security Robots, Artificial Intelligence, and Deadly Force." *UC Davis Law Review* 51, no. 2 (2017): 569–587.

Kania, Elsa B. "Chinese Military Innovation in Artificial Intelligence." *June 7 2019 Testimony before the U.S.-China Economic and Security Review Commission Hearing on Trade, Technology, and Military-Civil Fusion*. Washington: Center for a New American Security, 2019.

Keast, Jacinta. 2018. "Speaker Interview – Dr. Alessandro Arduino, Codirector of the Security and Crisis Management Program, Shanghai Academy of Social Science." *YCW: Young China Watchers*, March 24. http://www.youngchinawatchers.com/speaker-interview-dr-alessandro-arduino-co-director-of-the-security-and-crisis-management-program-shanghai-academy-of-social-science/.

Krahmann, Elke. *States, Citizens and the Privatisation of Security*. New York: Cambridge University Press, 2010.

Layton, Peter. 2017. "Commercial Drones: Privatising Air Power." *Lowy Institute: The Interpreter*, September 27. http://lowyinstitute.com.au/the-interpreter/commercial-drones-privitising-air-power

Layton, Peter. *Algorithm Warfare: Applying Artificial Intelligence to Warfighting*. Canberra: Air Power Development Centre, 2018.

Layton, Peter. "Robot Wars: UGVs on the Battlefield." *Defence Today* 12, no. 5 (2018): 13–17.

Legarda, Helena, and Marie L. Hoffmann. 2018. "China as a Conflict Mediator: Maintaining Stability along the Belt and Road." *MERICS: Mercator Institute for China Studies*, August 22. https://www.merics.org/en/china-mapping/china-conflict-mediator

Legarda, Helena, and Meia Nouwens. "Guardians of the Belt and Road: The Internationalization of China's Private Security Companies." In *MERICS China Monitor*, edited by Claudia Wessling and Ruth Kirchner. Berlin: MERICS: Mercator Institute for China Studies, (2018): 7, 11–12. Accessed August 6, 2018, 1–18.

Marten, Kimberly. "Russia's Use of Semi-State Security Forces: The Case of the Wagner Group." *Post-Soviet Affairs* 35, no. 3 (2019): 181–204. doi:10.1080/1060586X.2019.1591142.

Pilling, David. 2019. "The Fight to Control Africa's Digital Revolution." *Financial Times*, June 19. https://www.ft.com/content/24b8b7b2-9272-11e9-aea1-2b1d33ac3271

Plekhanov, Ilya. 2017. "Soldiers without Guns: How Do China PMC." *Center For Strategic Assessment And Forecasts*, July 17. http://csef.ru/en/oborona-i-bezopasnost/340/soldaty-bez-oruzhiya-kak-dejstvuyut-kitajskie-chvk-7783

PSCA. 2016. "Lahore to Get High Grade Surveillance Equipment under Safe City Project." *Punjab Safe Cities Authority*, September 3. https://psca.gop.pk/PSCA/2016/09/03/lahore-to-get-high-grade-surveillance-equipment-under-safe-city-project/

Ramachandran, Sudha. 2019. "Protecting BRI: China's Foreign Security Concerns." *The Central Asia-Caucasus Analyst*, March 14. https://www.cacianalyst.org/publications/analytical-articles/item/13560-protecting-bri-chinas-foreign-security-concerns.html

Reuters. 2017. "Security Firms to Cash in Protecting China's 'New Silk Road'." *South China Morning Post*, April 24. https://scmp.com/news/china/diplomacydefence/article/2090033/security-firms-cash-protecting-chinas-newsilk-road

Robinson, Gwen. 2019. "Southeast Asia Gains New Leverage as China and US Battle for Influence." *Nikkei Asian Review*, May 29. https://asia.nikkei.com/Spotlight/Cover-Story/Southeast-Asia-gains-new-leverage-as-China-and-US-battle-for-influence

Rossiter, Ash. "Drone Usage by Militant Groups: Explaining Variation in Adoption." *Defense and Security Analysis* 34, no. 2 (2018): 113–126. doi:10.1080/14751798.2018.1478183.

Rossiter, Ash. "Bots on the Ground: An Impending UGV Revolution in Military Affairs?" *Small Wars & Insurgencies* 31, no. 4 (2020).

Singer, P.W. *Corporate Warriors: The Rise of the Privatized Military Industry*. Ithaca: Cornell University Press, 2003.

Sponsored. 2018. "Here's How Punjab Police Integrated Command, Control and Communication Centre Is Making Lahore Safer." *ProPakistani*, June 19. https://propakistani.pk/2018/06/19/heres-how-punjab-police-integrated-command-control-and-communication-centre-is-making-lahore-safer/

Thiercelin, Arnaud. 2018. "Beyond Personal Drones: How the Largest Drone Manufacturer Transforms Itself." *AI Frontiers*, August 11. https://medium.com/aifrontiers/the-evolution-of-drones-how-the-largest-drone-manufacturer-transforms-itself-6b7e7a708f3f

Wang, Maya. 2019. "China's Algorithms of Repression: Reverse Engineering a Xinjiang Police Mass Surveillance App." *Human Rights Watch*, May 1. https://www.hrw.org/report/2019/05/01/chinas-algorithms-repression/reverse-engineering-xinjiang-police-mass-surveillance

Yang, Yingzhi. 2019. "Chinese AR Start-up Develops Smart Glasses to Help Police Catch Suspects." *South China Morning Post*, May 6. https://www.scmp.com/print/tech/start-ups/article/3008721/chinese-ar-start-develops-smart-glasses-help-police-catch-suspects

Yang, Zi. "Securing China's Belt and Road Initiative". *Special Report 436*. Washington: United States Institute of Peace, November 2018.

Zhou, Christina. 2019. "China's Belt and Road Forum Shows Xi Jinping's 'Genius Marketing Ploy' Has Become 'Project of the Century'." *Australian Broadcasting Commission News*, April 25. https://www.abc.net.au/news/2019-04-26/what-is-chinas-trillion-dollar-belt-and-road-initiative/11042036

The impact of Artificial Intelligence on hybrid warfare

Guilong Yan

ABSTRACT
Through a brief survey of the typical definitions of hybrid warfare (HW), this article illustrates the five salient features of HW: synergy, ambiguity, asymmetry, innovative disruption and battle over psychology; then based on a HW model proposed by Erik Reichborn-Kjennerud and Patrick Cullen, the article discusses the impact of Artificial Intelligence on the five instruments of power – military, political, economic, civil and informational (MPECI), and analyses the changes and continuities of HW in the age of Artificial Intelligence.

I. Introduction

Alex Roland, a historian at Duke University, holds that 'technology, more than any other outside force, shapes warfare.'[1] This observation is echoed by many scholars and practitioners of war worldwide. However, when proposing the concept of Hybrid Warfare (HW), James Mattis and Frank Hoffman emphatically downplayed the role of technology. In their view, the reason was not that technology is unimportant, but rather, that the obsession with technology by contemporary military professionals had blinded them to the predominance of human dimension in warfare.[2] Later, in giving his definition of 'hybrid threat', Hoffman intentionally eliminated 'disruptive technology' as a modality,[3] as opposed to the original description by Nathan Freier, who regards the future challenges in the security environment as a 'quad chart' of 'irregular, traditional, catastrophic and disruptive' threats that exploit revolutionary technology to negate the US military superiority.[4] Fortunately, Hoffman rectified his view on HW in later writings, as he stated, 'The term "hybrid" reflects more than a cross-breeding or blurring of regular and irregular tactics ... the fusion of advanced capabilities with the fluidity of irregular tactics is key and has been borne out repeatedly over the past decade.'[5]

The rapid development in Artificial Intelligence (AI) provides new advanced capabilities in both military and non-military domains, and this will not only transform the hardware of war but also impact the human dimension by direct or indirect means. Arguably, AI is a most disruptive technology in the current security environment, and the combination of AI with 'the fluidity of irregular tactics' will soon render the current understanding of HW obsolete. It is changing and will change the character of HW.

The past 2 years have seen an increased interest in research on AI and its impact on national security, defence, and even the future of warfare,[6] which yields useful insight into these areas and helps inform the related discourse. However, a focused research on the impact of AI on HW is still wanting. This article aims to probe into this niche, and hopefully generate debate on future research. It will plough through representative definitions of HW, and outline the defining characteristics of HW, which will serve as criteria for assessing AI's impact. An HW model is adopted in the analysis, and a focused discussion on AI's impact on the five instruments of power in HW is presented, and finally, conclusion is drawn based on the analysis.

II. Hybrid warfare: an evolving concept

For over a decade, HW has become a popular yet controversial term in the academic and military discourse. Its popularity lies mainly in capturing the form of struggle which features a combination of new technologies and fanatic fighting styles by non-state actors, and later the 'covert' operations of state actors that use deniable or paramilitary forces and incremental tactics to achieve the political aim without triggering the response of war. However, critics quickly point out that HW is not new, as numerous historical cases illustrate even in conventional wars there was plenty of irregular activity.[7] Besides, there is conceptual ambiguity and intellectual incoherence in the term, since different commentators define HW in different ways.[8]

The variations in definition are partly because the concept is deduced from people's observation of the enemy. As the enemy adapts to the changing environment and adopts new ways of fighting, so the observers find new dimensions and characteristics that used to be opaque or less visible.

While HW does not constitute a new form of warfare, the term caught on and took hold. One explanation is that HW is novel and useful in describing the changing character of war, as 'the differences in the equipment, weapons, training, and skills between contemporary regular and irregular forces are significantly bigger than they used to be in the past, and therefore their mix creates a truly new tactical environment.'[9] The other explanation is that it reflects the West's perceptions of the weakness and decline of itself, and its perceptions of the strength and guile of the adversaries.[10] In other words, HW reflects the West's anxiety over the loss of operational advantages brought by its technological edge.

When Hoffman first put forward the HW concept, his reference was the 9/11 terrorist attack and the two wars. Therefore, he described Hybrid Threats as 'threats that incorporate a full range of different modes of warfare including conventional capabilities, irregular tactics and formations, terrorist acts including indiscriminate violence and coercion, and criminal disorder, conducted by both states and a variety of non-state actors.'[11] Here his emphasis is laid on the incorporation of various 'modes of warfare' by state and non-state actors, and the salient feature is the combination of conventional capabilities with irregular tactics and even acts of terrorism.

McCuen was more aware of the expansion of battleground in HW beyond traditional thinking, as he put it, 'the decisive battles in today's hybrid wars are fought not on conventional battlegrounds, but on asymmetric battlegrounds within the conflict zone population, the home front population, and the international community population.'[12] He professed that battles fought within these populations ultimately determined success or failure. Therefore, he defined HW as 'a combination of symmetric and asymmetric war in which intervening forces conduct traditional military operations against enemy military forces and targets while they must simultaneously – and more decisively – attempt to achieve control of the combat zone's indigenous populations by securing and stabilizing them (stability operations).'[13]

Still, there are others who take a more expansive view of HW and regard it as just a continuation of historical cases. They define HW as 'conflict involving a combination of conventional military forces and irregulars (guerrillas, insurgents, and terrorists), which could include both state and non-state actors, aimed at achieving a common political purpose.'[14] This definition is criticized for ignoring the non-military means and the societal front, which should feature prominently in the current discourse of HW.[15]

After the beginning of the Ukrainian Crisis in 2014, the idea of HW allegedly went through a 'reconceptualization process', when the Russian actions in Crimea and eastern Ukraine displayed a mix of military and non-military means, and leveraged both conventional and irregular components, plus other instruments of power like cyber and information operations. This came as a shock to the Western community and quickly gave rise to discussions of 'Russian HW'. According to Ofer Fridman, the idea of 'Russian HW' was politicised by NATO and served as a panacea to its identity crisis.[16]

Therefore, the term HW took on a life of its own after Hoffman, scholars came up with numerous definitions, which cannot possibly be exhausted here. Even Hoffman himself laments that HW, as understood now, is quite different from what he and General Mattis put forward.[17] Suffice it to say, the proliferation of definitions does not contribute to a consensus among scholars, let alone military professionals and practitioners, on what HW is. As a GAO report shows, there were diverging views over the term HW among different entities within DOD,[18] so that Pentagon did not bother to come up with an official definition.

However, despite all the inadequacies and controversies, HW is a useful concept to describe current and future security challenges, and is 'a valuable way of describing the intellectual challenges adversaries are bringing to the table in terms of what war is and how it should be understood.'[19] It highlights the salient characteristics of this form of warfare that otherwise are obscured or neglected by contemporary defence analysts and military professionals.

III. Characteristics of hybrid warfare

Many scholars and military professionals have tried to describe the salient features of HW, but as their subjects of analysis range from non-state to state actors, and the particular dimension(s) they emphasize in their discourse change(s), their descriptions show much divergence rather than convergence. In addition, as hybrid threats and the modes and means of HW evolve, more salient characteristics will surface, which add to or even obscure those features that were originally held to be the litmus test.

In summarizing the characteristics of HW here, the author also runs the risk of being partial and biased in judgment; however, with the benefits of hindsight built on the observation of the empirical evidence in recent years and the copious literature on HW, it is possible to come up with a clearer depiction of the distinguishing characteristics of HW.

The first characteristic of HW lies in its synergistic use of various means across the military, political, economic, civilian and informational (MPECI) spectrum to exploit the vulnerabilities of the enemy. Here synergy is the key because the whole is greater than the sum of its parts. The combinational use of military and non-military means is the prominent feature that gives HW its name. Based on its available capability, political goals, and enemy weaknesses, the hybrid actor will use all means possible to achieve the desired effects, mostly targeting more heavily on the political and psychological domains. It may calculate and choose to escalate vertically or horizontally across the five dimensions (MPECI), so as to manipulate the process and intensity of HW, with the aim of achieving the best result.

The second major element of HW is its ambiguity. Ambiguity may include the blurred boundary between peace and war. While it is intuitive to view peace and war as inherently opposite categories, HW makes the distinction ambiguous, as it features activities or operations that defy easy categorization. Invariably, HW is designed to fall below the threshold of war and to delegitimize (or even render politically irrational) the ability to respond by military force.[20] Similarly, when this applies to non-state actors, ambiguity may also mean the fuzziness between organized violence, terrorism, criminal behaviour, and war. In addition, ambiguity may also refer to the obscure identity of combatants in war, as states may deploy troops without insignia and use proxies to hide agency; sometimes civilians are just warfighters in disguise.

Asymmetry is the third characteristic in HW. The capacity between state and non-state actors is intrinsically unequal. For example, in the military domain, states have well-trained troops and enjoy a great advantage in high-tech weaponry; in contrast, non-state actors mainly rely on a ragtag army equipped with low-end weapons. However, advantage in the asymmetry is relative, as non-state actors can launch attacks by illegal, criminal, and terrorist means, while despite a few exceptions, state actors are constrained by legal norms in counterattacks; sometimes non-state actors are armed with violent extremist ideology, which enables them to fight more fanatically, while state actors are duty-bound by ethics in war. And also because of the disparity in the strengths between state and non-state actors, a minor gain by the latter can be amplified through the media, often causing asymmetric damage to the former.

The fourth characteristic is the disruptive innovation at the tactical or operational levels, which helps to achieve strategic surprise. For non-state actors, disruptive innovation is achieved by repeated use of suicide attacks, adaptive use of modern commercial technology for military purposes, and is also demonstrated by their increased levels of military sophistication, like wielding modern weapon systems (anti-ship missiles and UAVs), secure communication, and sophisticated command and control, which were thought to be the exclusive possessions of the state. For state actors, they do so by rapidly mobilizing and inserting non-attributable forces, disinformation campaigns in media and the cyber domain, like the secrecy and suddenness of Russian military actions in Eastern Europe that took American intelligence by surprise.[21]

The fifth salient feature is the battle over the psychology of the target population. In this sense, HW emphasizes less on kinetic clashes between warring parties, but more on a contention over the narrative, perception and moral high ground. This is the reason why McCuen attaches such importance to the conflict zone population, the home front population, and the international community population. Those who can convince these populations of the righteousness of their cause, the totality of their victory, and the astounding defeat of the enemy have a better chance of 'winning' the HW. Therefore, state actors tend to threaten but economize the use of military force, so as to maximize the psychological pressure on the target population, while non-state actors tend to concentrate their efforts on striking a deadly blow at the enemy, and then magnify their success and undermine enemy morale through propaganda and (dis)information campaigns. This is of vital importance because success or failure in HW hinges less on military gains or losses, but more on the perceived victory or defeat.

IV. An HW model

Two scholars created an HW model (Figure 1) in order to provide conceptual clarity to HW by putting state and non-state actors under a single analytical framework. The model shows the dynamics of HW as a synchronized horizontal and vertical escalation along five major dimensions (MPECI: military, political, economic, civilian and informational) to achieve the desired goals. This model best captures the dynamic and elastic nature of HW in these dimensions.

According to the designers of the model, 'HW is best understood by focusing on various characteristics of the actors' capabilities and vulnerabilities as well as the ways the means are employed and to what effects. This unity of hybrid warfare is built on a number of characteristics: *it is asymmetric and multi-modal along a horizontal and a vertical axis, and to varying degrees shares an increased emphasis on creativity, ambiguity, and the cognitive elements of war.*'[22] And they further emphasize the importance of synchronization in vertical and horizontal escalation, the non-linear nature and thus the blurring of 'phases' in HW, and the expansion of battlefield across the PMESII (political, military, economic, societal, informational, and infrastructure) spectrum.[23]

For the purpose of this article, the utility of this model lies not only in depicting some of the core characteristics of HW, but more importantly, in delineating five major instruments of power that HW actors can employ and exploit. The five categories of power instruments, although broad, provide a feasible framework for the analysis of AI's impact on HW. Following this reasoning, and in order to analyze the impact of AI on HW, we first need to diagnose how AI is affecting each instrument of power, how it is adding to the strengths or creating vulnerabilities for the actors in HW, and then based on the analysis of each dimension, we should draw a holistic picture and see how a synchronized employment of these instruments of power can be escalated or deescalated to achieve effects. In analyzing the overall picture, it is important to always bear in mind that the whole is greater than the sum of its parts.

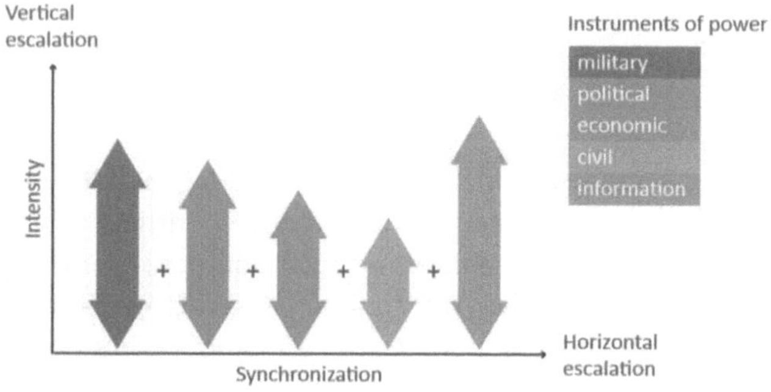

Figure 1. Hybrid warfare model.
Cullen and Reichborn-Kjennerud, "Understanding Hybrid Warfare," 9.

V. AI and the five instruments of power (MPECI)

The definition of Artificial Intelligence also varies considerably, as illustrated by Stephan De Spiegeleire et al., most concrete definitions of AI fall into four categories, depending on a combination of factors such as the emphasis on the attainment of thought processes or goal-oriented behavior, and the measurement of success against human performance or 'rationality'.[24] This article does not aim to delve into the technicalities of the definition, but try to give a holistic picture of how the application of AI may impact HW. In lieu of this, I choose a succinct definition of AI as 'the use of digital technology to create systems that are capable of performing tasks commonly thought to require intelligence'.[25]

When thinking about the sequence and stage of AI development, people generally classify AI into three tiers: Artificial Narrow Intelligence (ANI), Artificial General Intelligence (AGI), and Artificial Superintelligence (ASI). ANI refers to machine intelligence that equals or exceeds human intelligence for specific tasks; AGI refers to machine intelligence meeting the full range of human performance across any task; and ASI refers to machine intelligence that exceeds human intelligence across any task.[26] The author intends to build the analysis on the foundation of empirical evidence and hard reasoning, so as to depict a plausible scenario of HW at the current stage of AI development. Therefore, it is important to note that this article will only consider AI technologies that are currently available, i.e. ANI, and what HW will be like at stages of AGI and ASI is beyond the scope of this article.

Next, I would follow the guidelines of the HW model, and explore how AI can be utilized in the five instruments of power: military, political, economic, civil and informational (MPECI), and what impact this might have on the character of HW.

1. Military dimension

Recent years have seen a vast explosion in the number and variety of AI-enabled military systems, ranging from ground, aquatic to aerial robots. Although there are no such autonomous systems in operation that can demonstrate an independent capacity for knowledge- and expert-based reasoning, the trend is set to enable higher autonomy for a variety of military tasks.

Incremental progress has been made in many countries to make ground, air, on-water and underwater vehicles more autonomous. For example, Israel revealed that it had deployed fully autonomous robots – self-driving military vehicles – to patrol the border along Gaza Strip, and it planned to equip the vehicles with weapons, and deploy them in stages to Israel's frontiers with Egypt, Jordan, Syria, and Lebanon.[27] As for UAVs, there were at least 150 different military drone systems being used by 48 countries by September

2017, based on the database collected by the Centre for a New American Security and Bard College.[28] They range in size from the hummingbird size Black Hornet mini-copter to the massive 15,000-pound RQ-4 Global Hawk. Notably, Pentagon's Strategic Capabilities Office tested the Perdix nano drones in a swarm in 2017, when 100 Perdix drones were dropped from an F-18 'mother aircraft', autonomously formed a swarm, and did a series of manoeuvres successfully.[29] When it comes to aquatic robots, the Seahunter, a drone ship designed for anti-submarine warfare, just joined the US navy in February 2018, after a two-year testing and evaluating program.[30] It is designed to autonomously carry out 70 daylong sea surface patrols at a time, as far out from base as 10,000 nautical miles.[31] And the list of AI-enabled military systems goes on and on.

Yet autonomous weaponry is only part of the picture, AI has shown immense utility and potential in military C4ISR. Take Command and Control (C2) as an example, AI can better deal with information overload, improve situational awareness and recommend courses of action, as described by Ayoub and Paine, 'A modular AI that can optimize some tactical activity – say, the storming of an enemy position, rapidly coordinating fires and manoeuvre via networked and automated platforms – would outperform a seasoned battalion commander with ease.'[32] The US Marine Corps intends to acquire an AI-based Command and Control digital assistant, which 'uses advanced computing techniques such as machine learning and natural language processing to provide answers to complex mission-specific questions to enhance battlespace decision-making.[33] In addition, AI is also used in developing rapid and accurate automatic target recognition (ATR) systems, like the ATR program prototype developed by Deep Learning Analytics for DARPA, to trial systems assisting pilots in finding and engaging targets.[34] As for intelligence, Ash Carter, former US deputy secretary of defense, said in 2015 that autonomous deep learning machine systems are able 'to see the patterns of hybrid warfare, to give early warning that something is happening in gray zone conflict areas, and to respond at extreme speed, and under rapidly shrinking engagement windows.'[35] Actually, some companies are making predictions based on machine learning algorithms to do an analysis of big online data, e.g. Predata, a US consulting firm, claims to 'use anonymized online metadata to predict geopolitical, security, and market-moving events'. According to its co-founder James Shinn, it is able to predict events 1–10 days ahead.[36] When it comes to surveillance and reconnaissance, experts claim that more autonomous and survivable drones could potentially lead to greater stability between states by enhancing the monitoring of contested areas, reducing the viability of covert or hybrid operations.[37]

Currently, it seems state militaries have a monopoly on the most sophisticated weaponized drones and AI-enabled C4ISR capabilities, however, this does not mean that state-actors are assured of an asymmetrical advantage in

HW. The reasons are as follows: First of all, as commercial companies worldwide pour more money into the R&D of AI, over time, there will be a fundamental shift in technology prowess whereby militaries will start to significantly lag in autonomous system capabilities as compared with commercial systems, and terrorists will potentially be able to buy drones on the internet with as much or greater capability than those available to the military.[38] Besides, 'the gap between historical military superiority in UAV development and the present capabilities of the commercial sector is closing, as evidenced by the increasing number of military-grade drones offered for sale via the internet.'[39] Secondly, even equipped with inferior or commercially available drones, non-state actors and especially terrorists can turn them into lethal weapons by loading explosives in the drones. There have been at least two reports of deliberate drone attacks by terrorists this year, one was a swarm of low-cost drones armed with high-explosives attacking the Russian Khmeimim airbase and Russian Naval CSS point in the city of Tartus, Syria;[40] the other was an attack on a Saudi oil tanker near the Red Sea by Houthi rebels using three drone boats armed with explosives.[41] Although both attacks were thwarted, the dangers of using AI-enabled IEDs by the technically and materially weak parties were highlighted. Thirdly, the AI-enabled C4ISR cannot help dissipate the fog of war. As is eloquently argued by Rodrick Wallace, 'Cognitive algorithmic entities tasked with the real-time management of critical processes under rapidly shifting "roadway" conditions will face many of the same conundrums and constraints that confront the conduct of warfare and other forms of conflict.'[42] Therefore, surprise and uncertainty are still prominent features of HW, even in the age of AI.

2. *Political dimension*

AI-enabled tools have allegedly been used to conduct disinformation campaigns by Russia as part of its political warfare against the West. The tools include automated accounts (bots) and impersonation accounts on social media, and the goals are to undermine the Western political narrative and trans-Atlantic institutions, sow discord and divisions within countries, and blur the line between fact and fiction. The methods Russia uses include full-spectrum dissemination and amplification of misleading, false, and divisive content, and deployment of computational propaganda.[43] It is said that 'between May and July 2019, bots accounted for a 55 percent of all Russian-language Twitter messages on NATO presence in the Baltic states and Poland,' and that Russian use of bots has shown sharp changes in tactics over time, ranging from 'post-bots' to 'news-bots' and 'mention-trolls', reflecting growing sophistication and detection-proof measures along the way.[44]

It is also well known that machine intelligence has been employed in election campaigns to influence voters and manipulate public opinion.

During the 2016 US presidential election, the data science firm Cambridge Analytica rolled out an extensive advertising campaign to persuade voters based on their individual psychology; massive swarms of political bots were used in the 2017 general election in the UK to spread misinformation and fake news on social media.[45] The AI-enabled campaign 'aids' can automatically generate a tailored political message to target individual voters, based on their political preference and personal likings, and this is an increasingly widespread tactic that attempts to shape public discourse and distort political sentiment.

In addition, AI technology can help forge video and audio with ease, which poses an enormous challenge to the article of faith – 'Seeing is believing'. This can be leveraged to achieve political effects. A particular worrisome prospect is that the technology is proliferating. For example, an AI company called Lyrebird is developing technology that allows anyone to produce surprisingly realistic-sounding speech with the voice of any individual. Lyrebird's demo generates speech, including varied intonation, in the voices of Donald Trump, Barack Obama, and Hillary Clinton.[46] Another system jointly developed by the University of Erlangen-Nuremberg, Max-Planck-Institute for Informatics, and Stanford University allows an individual to control the facial expressions of someone in a video, and the demo shows how they use Bush, Trump, Putin, and Obama as the 'target actors' and effectively controls their expressions.[47] Therefore, AI can generate high-quality fake video footage of influential politicians, by manipulating the content, like making politicians 'say' appalling things, it can wreak havoc on the media, and potentially create a political crisis.

Although currently state actors seem to enjoy a huge advantage in employing the AI-enabled tools in creating targeted propaganda and manipulating videos, non-state actors soon will have access to these tools via commercial means. In the future, HW conducted in the political arena will be a contest over the speed of verifying and discrediting narratives, which will make the reality more volatile and perplexing.

3. Economic dimension

When HW is targeted at a state, economic coercion is often used as a means of leverage, as Russia did in its dealing with Ukraine in 2014. Since economic interdependence is the norm in today's globalized world, economic relationships are inherently susceptible to political manipulation for strategic purposes, from extending influence to exerting pressure, both overtly and covertly.[48] When HW is targeted at a non-state actor, financial cutoff and economic isolation would be a normal choice for states. In both cases, AI can play a role in tracking the economic, trade and financial activities of the target actor, and in coming up with refined analysis and optimized plans in punishing the target.

The adoption of AI applications in the economic and financial sectors has been underway, driven by both supply and demand factors. AI is used in various areas such as assessing credit quality, optimizing capital investment, trading transaction, regulatory compliance, and fraud detection.[49] While integrating AI into the economic sector brings huge benefits, it is also embedded with potential vulnerabilities. When it comes to non-state actors against state actors in HW, they may choose to launch either physical attacks on the economic infrastructure or cyber attacks on the financial hubs and networks, or both. The former would cause economic disorder and disruption, and the later would cause financial breakdown or even social chaos. The weapons the attackers use might not necessarily be AI-enabled for the physical attacks, and they may also choose AI-enabled cyber attacks on financial networks.

4. *Civil dimension*

Events like the Arab Spring illustrate how social media can act as a catalyst for popular movements to degenerate into social unrest and even political upheaval in societies that suffer from chronic corruption, unemployment, aging dictatorships and social discontent. The advent of AI has offered new tools of digital propaganda and disinformation to influence the civil society.

By exploiting the religious, sectarian and ethic divisions in the society, HW actors can automate realistic video and audio with the help of AI. They may fabricate sensational, seditious and inflammatory videos and audios by impersonating the national leadership or other prominent shapers and influencers, so as to sow discord among the enemy population. For example, with the help of AI tools, American comedian Jordan Peele produced a fake video of former President Obama, making him mimic whatever the manipulator says.[50] Commercial companies are making such technologies easily accessible online. Hybrid actors may even fabricate video or audio clips of the enemy leader reading out the Instrument of Surrender, so as to undermine the enemy morale and will to fight. The utility of AI can add to the scale and speed of impact. With good timing and seasoned manipulating skills, they may successfully turn a civilian unrest into a political movement, with devastating consequences for the enemy.

Likewise, the HW actors can also produce fake videos of catastrophic accidents or terrorist attacks to create panic among the populace, and then leverage it to achieve political effects. In so doing, they may choose to build on a real news report of major events, and then distort the truth by fabricating disastrous scenes like terrorist attacks or massive explosions. As early as in 2017, Nvidia demonstrated the ability of one of their AIs to generate disturbingly realistic videos of completely fake people, and it could do a surprisingly decent job of changing day into night, winter into summer, and house cats into cheetahs (and vice versa).[51]

5. *Informational dimension*

Arguably AI can have the greatest utility in the information domain. As an instrument of power, information can be leveraged to shape the political discourse, influence people's perception, and even change the political outcome. In an HW context, AI can be used in the following ways to exert influence.

First, fake news reports with realistic fabricated video and audio can be generated with the help of AI, which can be leveraged in various ways to achieve instantaneous and short-term effects, like creating shock and awe, causing panic, and disorder. But depending on the scale, intensity and sustainability of operation, the effects are unlikely to sustain for long, as the target audience may eventually find out the truth. But even of short duration, damage may have been done and purpose achieved by the time the momentum dies away.

Second, AI may also enable denial-of-information attacks with bot-driven, large-scale information flooding attacks. The purpose is to swamp information channels with noise, making it more difficult to acquire the real information.[52] But this should be tailored to major information outlets or channels in order to achieve the best effects, and may not sustain for a long duration, as the target network may detect intrusion and take defensive measures over time.

Third, AI can be used to manipulate information availability. It is no secret that the algorithms of search engines are designed to turn out personalized search results for the users. For example, even if two users search the same keyword on Google, there are distinguishable differences in the results they get. Similarly, media platforms' content curation algorithms are used to drive users towards or away from certain content in ways to manipulate user behaviour.[53] As the bias is embedded in the algorithm and not easily detectable, this can be a useful tool to manipulate information for long-term effects.

Of course, if we expand the concept of information to include the cyber domain, AI can be used either to find and attack the target network's vulnerabilities, or to mend and defend one's own. In 2016, DARPA hosted the Cyber Grand Challenge contest, in which rival teams competed with each other to create programs that could autonomously attack other systems while defending themselves.[54]

VI. HW in the age of narrow AI

When piecing all these dimensions together, we can tentatively depict a picture of HW in the age of narrow AI. It should be reminded that all the norms of HW as we saw in the past still persist in the future HW, the inclusion of AI in the contour will not render old ways of fighting obsolete, as the word 'hybrid' itself implies, there is a full spectrum of instruments at play in an HW, both low-end techniques like using IEDs, suicide bombers, cloak and dagger, and AI-enabled high-end tools will be employed. The advent of AI can neither help dissipate the 'fog of war', nor change the nature of war. That said, the inclusion of AI in HW does change the character of the war, as the military technological breakthroughs did in history.

In the military dimension, autonomous weapons will further free human beings from the kill chain, which may offer a more detached feeling to operators in the control of drones. Targeted killing will be more precise, as explosive-laden mini-drones combined with facial recognition technology may make pin-point elimination of enemy leaders easier. There may be a lopsided tactical victory for the militarily strong against the weak, especially when they engage in a fight of drones against the flesh. But this is not enough to ensure a strategic success, as the weak continues to 'have a vote' in the fight. By adapting and weaponising commercially available AI technologies and products, the weak has the potential of inflicting major damage on the opponents, and then it can amplify those gains by means of propaganda to gain strategic effects. This conforms to the observation that 'The integration of sophisticated robotics and automated tools may increase the differential between advanced Western forces and their challengers, which, in turn, will push state and non-state armed actors to develop more clandestine, unconventional and illegal methods such as the ones currently described as hybrid or "unrestricted".'[55]

In the political dimension, AI offers a more autonomous, more massive, and sometimes more covert means in carrying out propaganda and psychological warfare, with the aim of manipulating the political dynamic and discourse, like subverting the government, sowing discord, and creating division. This usually takes greater effect when combined with other instruments of power, like military coercion, economic leverage, and information campaigns.

In the economic dimension, AI can be used in diagnosing the critical vulnerabilities in the economic structure of the enemy, such as endowment of resources, volume of trade, and financial condition, and come up with calculated penalties against the enemy.

In the civil dimension, AI offers new tools of propaganda and disinformation to influence the civilian population. These tools can be capitalized on to exploit the contradictions and divisions in the society, and eventually to make political gains.

In the informational dimension, AI can help generate fake news reports, carry out denial-of-information attacks, and manipulate information availability. It can also be used in cyberattack and defence. These methods can be selectively used to achieve short-term or long-term gains.

Therefore, HW actors synchronize the use of the above five instruments of power, and depending on their goals, choose to vertically and horizontally escalate or deescalate across the spectrum as opportunities and timing allow. The kinetic aspect of war will become more lethal, precise and fast-tempo, when autonomous or semi-autonomous weapons are employed; and on the non-kinetic fronts of the war, the warring parties will put a premium on the acquisition, protection, analysis, and exploitation of data and information, because all five instruments of power in HW hinge on data and information. Wars are fought not only on the conventional and population battlegrounds but also on the virtual digital battleground; eventually, the outcome of HW will lean heavily towards those who command informational dominance.

VII. A caveat: AI, doctrine and organization

Historically, technological breakthroughs usually transformed the way we wage war and heralded the dawning of a new military era. Recent years have seen the astonishing speed of AI development, and its application in both military and non-military domains has also shown great momentum. It seems the AI is poised to revolutionize the old ways of warfighting. But the question is, will the advent of AI revolutionize the way we fight HWs?

Andrew Krepinevich's observation of the military-technical revolution may shed light on the question. According to him, 'Historical examples of past military-technical revolutions make clear that technological change by itself is insufficient to bring about a military-technical revolution. Innovative operational concepts and organizational innovations designed to exploit new technologies are crucial to a military's ability to realize large gains in military effectiveness.'[56] Therefore, although the advent of AI has brought enormous technological change, in order to revolutionize HW, we still need innovative operational concepts and organizational innovations.

Unfortunately, an innovative operational concept about HW is still wanting. As Cullen and Kjennerud point out, 'the international consensus on "hybrid warfare" is clear: no one understands it, but everyone, including NATO and the European Union, agrees it is a problem.'[57] However, it is encouraging that their research is a step forward in that they come up with an analytical framework of HW, and an HW operational concept might build on this, incorporating the five instruments of power and the changing technological know-how like AI. The advancement in AI research, development and applications offers opportunities for operational innovation, and hopefully, new concept might arise out of this.

As for organizational innovation, obviously, the current defence organizational structure lags far behind technological development. The military is notorious for its reluctance to make organizational adaptation or changes. But without organizational adaption, the defence sector will not benefit from the advantages brought by the advent of the AI era. In this regard, the structure of successful AI companies may give us some insight. Andrew Ng, chief scientist of Baidu, posits that building a centralized AI organization and matrix into the various business units is key to help transform the way companies do business, and he further emphasizes that strategic data acquisition, unified data warehouse, pervasive automation, and new job description are prerequisites for an AI company.[58] Therefore, for the defence department, it should consider incorporating a centralized AI unit into the department, making it the hub connecting the defence secretary and other units in the defence department.

Another model proposed by Spiegeleire et al. is also illuminating. They posit that 'the advent of AI requires us to look more broadly, probably quite differently, but above all more creatively – at "defence" and "armed force" and at the role AI is likely to play in those.'[59] They further propose a four-tiered classification of 'defence' as a social technology: Defence as military operators – the Defence Force; Defence as an organization that supports these operators but also interacts with its counterparts – the Defence Organization; Defence as a player in an increasingly more whole-of-government security-oriented approach – Defence and Security Organizations; and Defence as the (potential) catalyst of a broader defence and security ecosystem of sensors and effectors – the Defence and Security Ecosystem.[60] These four tiers form concentric circles, with the Defence Force at the centre and the other tiers successively outward. This broad and organic view of defence helps to mobilize all elements of government power in dealing with security threats, including hybrid threats.

VIII. Conclusion

It is interesting that cognitively we tend to 'remember' the future instead of creatively dreaming it up, which means we are 'stuck' in a particular mental frame that prevents us from thinking about truly different ways of doing things.[61] Therefore, the analysis of how AI will impact future HW is inevitably constrained by the author's mental frame, which is shaped by past events. In the future HW, innovative and fresh-new means of fighting enabled by AI will surely come out and will be used to achieve tactical or even operational surprise. However, the limitations of human imagination do not prevent us from drawing the conclusion that HW in the AI age will look very different that the kinetic aspect of war will become more lethal, precise and fast-tempo, and that the competition for control and dominance of data and information will be of vital importance.

Nevertheless, that does not mean the salient characteristics of HW will change. Features like synergistic use of various means, ambiguity, asymmetry, disruptive innovation and the battle over psychology still remain. Therefore, success might not be in the hands of those who are technologically powerful, the fog of war and friction will continue to plague parties engaged in an HW. This judgment is echoed by the following observation: 'One regular assumption was that the odds of success might be shifted decisively as a result of some new technology ... But the technology was rarely monopolized or else, even if one side enjoyed superiority, adversaries found ways to limit their effects. Even for modern Western forces, technology encouraged a fantasy of a war that was fast, easy, and decisive: yet they still found themselves facing "slow, bitter and indecisive war"'.[62]

In order to better cope with the challenges of HW in the age of AI, we need a fuller understanding of HW, and even an operational doctrine that incorporates AI as a salient component. In addition, innovative ideas about defense organization and structure should be encouraged, developed and executed, and this would contribute to building a more resilient, flexible, and intelligent defense organization, or even a Defense and Security Ecosystem as envisioned by scholars. Technological change and military systems evolution has gradually become a reality with the advent of AI, combined with operational innovation and organizational adaptation, we will revolutionize the way we fight HW.

Notes

1. Roland, "War and Technology."
2. Mattis and Hoffman, "Future Warfare," 56.
3. Instead of disruptive technology, Hoffman regards criminality as the fourth mode of hybrid threat, which reflects his emphasis on non-state actors as a more likely source of hybrid threats. Hoffman, "Hybrid vs. Compound War."
4. Freier, *Strategic Competition*, 2.
5. Hoffman, "The Contemporary Spectrum of Conflict," 29.
6. Allen and Chan, *Artificial Intelligence*; Spiegeleire, Maas and Sweijs, *Artificial Intelligence*; Cummings, "Artificial Intelligence."
7. Murray and Mansoor, eds., *Hybrid Warfare*.
8. Fridman, "The Danger of 'Russian Hybrid Warfare'"; and Tuck, "Hybrid War."
9. Fridman, "The Danger of 'Russian Hybrid Warfare'."
10. Tuck, "Hybrid War."
11. Hoffman, *Conflict in the 21st Century*, 8.
12. McCuen, "Hybrid Wars," 107.
13. Ibid, 108.
14. Mansoor, "Introduction," 2.
15. Tienhoven, "Identifying 'Hybrid Warfare'."
16. See note 9 above.
17. Senate Armed Forces Committee Hearing, "The Evolution of Hybrid Warfare."
18. U.S. Government Accountability Office (GAO), "National Defense".
19. Reichborn-Kjennerud and Cullen, "What is Hybrid Warfare," 1.
20. Reichborn-Kjennerud and Cullen, "What is Hybrid Warfare," 2.
21. Batyuk, "The US Concept and Practice of Hybrid Warfare," 468.
22. See note 21 above.
23. Ibid., 3.
24. Spiegeleire, Maas and Sweijs, *Artificial Intelligence*, 28–29.
25. Brundage, et al., "The Malicious Use of Artificial Intelligence," 9.
26. Spiegeleire, Maas and Sweijs, *Artificial Intelligence*, 30.
27. Ibid., 80.

28. Walker, "Unmanned Aerial Vehicles (UAVs)."
29. Dyndal, Berntsen and Redse-Johansen, "Autonomous Military Drones."
30. Austin, "Sea Hunter."
31. Miltner, "Can the U.S. Navy Brave the Waves of Autonomous Warfare?"
32. Ayoub and Payne, "Strategy in the Age of Artificial Intelligence," 806.
33. Kent, "Artificial Intelligence (AI)-based C2 Digital Assistant."
34. Spiegeleire, Maas and Sweijs, *Artificial Intelligence*, 88.
35. Ibid., 85.
36. Shinn, "Predictive Analytics."
37. Spiegeleire, Maas and Sweijs, *Artificial Intelligence*, 90.
38. Cummings, "Artificial Intelligence," 11.
39. Ibid., 12.
40. Rogoway, "Russia Says January 5th Attack on its Syrian Air Base Was by a Swarm of Drones."
41. Reuters Staff, "Saudi-led Coalition Says Thwarts Houthi Attack on Oil Tanker."
42. Wallace, *Carl von Clausewitz*, vii.
43. Polyakova and Boyer, *The Future of Political Warfare*, 4.
44. Anonymous, "Bots on the Ground."
45. Polonski, "How Artificial Intelligence Conquered Democracy."
46. Allen, "AI Will Make Forging Anything Entirely Too Easy."
47. Price, "AI and CGI Will Transform Information Warfare."
48. Ducaru, "Framing NATO's Approach to Hybrid Warfare," 6.
49. Financial Stability Board, "Artificial Intelligence."
50. ABC News, "Star uses AI, President Obama in fake news PSA."
51. Leary, "An AI That Makes Fake Videos May Facilitate the End of Reality as We Know It."
52. Brundage, et al., "The Malicious Use of Artificial Intelligence," 29.
53. Ibid.
54. DARPA, "The World's First All-Machine Hacking Tournament."
55. Johnson, "Hybrid War," 145.
56. Krepinevich, *The Military-Technical Revolution*, 1.
57. Cullen and Reichborn-Kjennerud, "Understanding Hybrid Warfare," 3.
58. Ng, "The State of Artificial Intelligence."
59. Spiegeleire, Maas and Sweijs, *Artificial Intelligence*, 67.
60. Ibid., 67–68.
61. Ibid., 66.
62. Betz, *Carnage and Connectivity*, 5.

Acknowledgement

The author would like to thank CCW Centre at Oxford University, and especially Dr. Rob Johnson and Dr. Annette Idler for their guidance and help, thanks also go to Dr. Paul Rich and Mr. Tom Durell Young for their comments, and Mr. Anthony Blacer and Dr. Ash Rossiter for editing.

Disclosure statement

No potential conflict of interest was reported by the author.

Funding

This work was supported by Chevening Fellowships, the UK government's global awards scheme, funded by the Foreign and Commonwealth Office (FCO) and partner organisations.

Bibliography

ABC News. 2018. "Star uses AI, President Obama in fake news PSA." Accessed November 2019. https://abcnews.go.com/GMA/News/video/star-ai-president-obama-fake-news-psa-54550809.

Allen, Greg. 2017. "AI Will Make Forging Anything Entirely Too Easy." *Wired*, January. Accessed April 2018. https://www.wired.com/story/ai-will-make-forging-anything-entirely-too-easy/

Allen, Gregory C., and Taniel Chan. 2017."Artificial Intelligence and National Security." *Belfer Center for Science and International Affairs. Harvard Kennedy School*, July. Accessed April 2018. https://www.belfercenter.org/sites/default/files/files/publication/AI%20NatSec%20-%20final.pdf

Anonymous. 2019. "Bots on the Ground: Half of Russian tweets on NATO in Baltics and Poland come from bot-networks." Accessed November 2019. https://www.lrt.lt/en/news-in-english/19/1094490/bots-on-the-ground-half-of-russian-tweets-on-nato-in-baltics-and-poland-come-from-bot-networks.

Austin, Mark. 2018. "'Sea Hunter,' a Drone Ship with No Crew, Just Joined the U.S. Navy Fleet." February. Accessed April 2018. https://www.digitaltrends.com/cool-tech/darpa-sea-hunter-joins-navy-fleet/

Ayoub, Kareem, and Kenneth Payne. "Strategy in the Age of Artificial Intelligence." *Journal of Strategic Studies* 39, no. 5–6 (2016): 793–819.

Batyuk, Vladimir I. "The US Concept and Practice of Hybrid Warfare." *Strategic Analysis* 41 (2017): 5. doi:10.1080/09700161.2017.1343235.

Betz, David. *Carnage and Connectivity: Landmarks in the Decline of Conventional Military Power*. London: Hurst, 2015.

Brundage, Miles, Shahar Avin, Jack Clark, Helen Toner, Peter Eckersley, Ben Garfinkel, Allan Dafoe. 2018. "The Malicious Use of Artificial Intelligence: Forecasting, Prevention, and Mitigation." *Future of Humanity Institute*, et al., February. Accessed April 2018. https://maliciousaireport.com/

Cullen, Patrick., and Erik Reichborn-Kjennerud. 2017. "Understanding Hybrid Warfare." *Multinational Capability Development Campaign*, January. Accessed April 2018. https://assets.publishing.service.gov.uk/government/uploads/system/uploads/attachment_data/file/647776/dar_mcdc_hybrid_warfare.pdf

Cummings, Mary L. 2017. "Artificial Intelligence and the Future of Warfare." *Chatham House*. January. Accessed April 2018. https://www.chathamhouse.org/sites/files/chathamhouse/publications/research/2017-01-26-artificial-intelligence-future-warfare-cummings-final.pdf

DARPA. 2016. "The World's First All-Machine Hacking Tournament." Accessed April 2018. http://archive.darpa.mil/cybergrandchallenge/.

Ducaru, Sorin Dumitru. "Framing NATO's Approach to Hybrid Warfare." In *Countering Hybrid Threats: Lessons Learned From Ukraine*, edited by Niculae Iancu, Andrei Fortuna, and Cristian Barna, 6. Amsterdam: IOS Press, 2016.

Dyndal, Gjert Lage, Tor Arne Berntsen, and Sigrid Redse-Johansen. 2017. "Autonomous Military Drones: No Longer Science Fiction." July. Accessed April 2018. https://www.nato.int/docu/review/2017/also-in-2017/autonomous-military-drones-no-longer-science-fiction/EN/index.htm

Financial Stability Board. 2017. "Artificial Intelligence and Machine Learning in Financial Services." November. Accessed April 2018. http://www.fsb.org/wp-content/uploads/P011117.pdf

Freier, Nathan. *Strategic Competition and Resistance in the 21st Century: Irregular, Catastrophic, Traditional, and Hybrid Challenges in Context*. PA: United States Army War College, 2007.

Fridman, Ofer. "The Danger of 'Russian Hybrid Warfare'." Accessed September 2019. http://www.cicerofoundation.org/lectures/Ofer_Fridman_The_Danger_of_Russian_Hybrid_Warfare.pdf

Hoffman, Frank. "Hybrid Vs. Compound War, the Janus Choice: Defining Today's Multifaceted Conflict." *Armed Forces Journal* (October, 2009). Accessed April 2018. http://armedforcesjournal.com/hybrid-vs-compound-war/

Hoffman, Frank G. *Conflict in the 21st Century: The Rise of Hybrid Wars*. Arlington, VA: Potomac Institute for Policy Studies, 2007.

Hoffman, Frank G. "The Contemporary Spectrum of Conflict: Protracted, Gray Zone, Ambiguous, and Hybrid Modes of War." In *2016 Index of U.S. Military Strength: Assessing America's Ability to Provide for the Common Defence*, edited by Dakota L. Wood, 29. Washington, DC: The Heritage Foundation, 2015.

Johnson, Robert. "Hybrid War and Its Countermeasures: A Critique of the Literature." *Small Wars & Insurgencies* 29 (2018): 1. doi:10.1080/09592318.2018.1404771.

Kent, Jeffrey. "Artificial Intelligence (Ai)-based C2 Digital Assistant." Accessed April 2018. http://www.navysbir.com/n16_2/N162-074.htm

Krepinevich, Andrew F. *The Military-Technical Revolution: A Preliminary Assessment*. Washington, DC: CSBA, 2002.

Leary, Kyree. 2017. "An AI That Makes Fake Videos May Facilitate the End of Reality as We Know It." Accessed November 2019. https://futurism.com/ai-makes-fake-videos-facilitate-end-reality-know-it.

Mansoor, Peter R. "Introduction: Hybrid Warfare in History." In *Hybrid Warfare: Fighting Complex Opponents from the Ancient World to the Present*, edited by Williamson Murray and Peter R. Mansoor, 2. Cambridge: Cambridge University Press, 2012.

Mattis, James N., and Frank Hoffman. "Future Warfare: The Rise of Hybrid Wars." *U.S. Naval Institute Proceedings* 132, no. 11 (November, 2005): 1–2.

McCuen, John J. "Hybrid Wars." *Military Review* 88, no. 2 (March–April, 2008): 107–113.

Miltner, Olivia. 2018. "Can the U.S. Navy Brave the Waves of Autonomous Warfare?" May. Accessed April 2018. https://www.ozy.com/fast-forward/can-the-us-navy-brave-the-waves-of-autonomous-warfare/82418

Ng, Andrew. 2017. "The State of Artificial Intelligence." *The Artificial Intelligence Channel*, December. Accessed April 2018. https://www.youtube.com/watch?v=NKpuX_yzdYs

Polonski, Vyacheslav W. 2017. "How Artificial Intelligence Conquered Democracy." August. Accessed April 2018. https://www.independent.co.uk/news/long_reads/artificial-intelligence-democracy-elections-trump-brexit-clinton-a7883911.html

Polyakova, Alina, and Spencer P. Boyer. *The Future of Political Warfare: Russia, the West, and the Coming Age of Global Digital Competition*. Washington, DC: Brookings Institution, 2018.

Price, Rob. 2017. "AI and CGI Will Transform Information Warfare, Boost Hoaxes, and Escalate Revenge Porn." *Business Insider*, August. Accessed April 2018. http://uk.businessinsider.com/cgi-ai-fake-video-audio-news-hoaxes-information-warfare-revenge-porn-2017-8

Reichborn-Kjennerud, Erik, and Patrick Cullen. "What is Hybrid Warfare." *Policy Brief [1/2016]. Norwegian Institute of International Affairs*. Accessed April 2018. https://brage.bibsys.no/xmlui/bitstream/id/411369/NUPI_P

Reuters Staff. 2018. "Saudi-led Coalition Says Thwarts Houthi Attack on Oil Tanker." January 10. Accessed April 2018. https://www.reuters.com/article/us-shipping-redsea-attack/saudi-led-coalition-says-thwarts-houthi-attack-on-oil-tanker-idUSKBN1EZ2G8

Rogoway, Tyler. 2018. "Russia Says January 5th Attack on its Syrian Air Base Was by a Swarm of Drones." January 8. Accessed April 2018. http://www.thedrive.com/the-war-zone/17493/russia-says-january-5th-attack-on-its-syrian-air-base-was-by-a-swarm-of-drones

Roland, Alex. 2009. "War and Technology." February. Accessed April 2018. https://www.fpri.org/article/2009/02/war-and-technology/

Senate Armed Forces Committee Hearing. 2017. "The Evolution of Hybrid Warfare and Key Challenges." March 22. Accessed April 2018. https://www.youtube.com/watch?v=s4ais_PH4ic

Shinn, James. 2018. "Predictive Analytics and National Security." *Lecture delivered at Oxford University*, April 24.

Spiegeleire, Stephan De, Matthijs Maas, and Tim Sweijs. 2017. *Artificial Intelligence and the Future of Defense: Strategic Implications for Small- and Medium-sized Force Providers*. The Hague Centre for Strategic Studies. Accessed April 2018. https://hcss.nl/sites/default/files/files/reports/Artificial%20Intelligence%20and%20the%20Future%20of%20Defense.pdf

Tienhoven, Manon van. 2016. "Identifying 'Hybrid Warfare'." Master Thesis, University of Groningen. Accessed April 2018. https://openaccess.leidenuniv.nl/bitstream/handle/1887/53645/2016_Tienhoven_van_CSM.pdf?sequence=1

Tuck, Chris. 2017. "Hybrid War: The Perfect Enemy." April. Accessed September 2019. https://defenceindepth.co/2017/04/25/hybrid-war-the-perfect-enemy/

U.S. Government Accountability Office (GAO). 2010. "National Defense: Hybrid Warfare: GAO-10-1036R." September. Accessed April 2018. https://www.gao.gov/assets/100/97053.pdf

Walker, Jon. 2017. "Unmanned Aerial Vehicles (Uavs) – Comparing the USA, Israel, and China." September. Accessed April 2018. https://www.techemergence.com/unmanned-aerial-vehicles-uavs/

Wallace, Rodrick. *Carl Von Clausewitz, the Fog-Of-War, and the AI Revolution: The Real World Is Not A Game of Go*, vii. Cham: Springer, 2018.

Index

Note: Page numbers in **bold** refer to tables; page numbers in *italics* refer to figures.